AuthorHouse™
1663 Liberty Drive
Bloomington, IN 47403
www.authorhouse.com
Phone: 1-800-839-8640

Published by AuthorHouse 04/10/2013

ISBN: 978-1-4772-9653-0 (sc)
*   978-1-4817-4263-4 (e)*

Library of Congress Control Number: 2012923117

authorHOUSE®

Enrico Fermi (September 29, 1901 - November 28, 1954), Nobel Prize winner in Physics in 1938, known for his contributions to the development of quantum theory, nuclear physics, and particle physics, once said to his student Leon Lederman, he himself the 1988 Nobel Prize recipient in Physics for his work on neutrinos,

"Young man, if I could remember the names of these particles, I would have been a botanist!"

Laboratory Schedule

Phylogeny

Domain Eukarya
      Supergroup Amebozoa
            Phylum Myxomycota (Slime Molds)
                  Class Myxomycetes (Plasmodial)
                  Class Dictyosteliomycetes (Cellular)
                  Class Protosteliomycetes (Protostelids)

Related Terms: myxamoebas, phagocytosis, pseudoplasmodium, macrocyst, plasmodium, sporangia (advanced), aethelia, plasmodiocarp (primitive)

Domain Eukarya
      Supergroup Opisthokonta
            Phylum Chytridiomycota (Chytrids)

Related Terms: holocarpic (parasitic), eucarpic (parasitic), rhizomycelium (coenocytic - multinucleate), non-nucleate rhizoids, uniflagellate spores, zoosporangia, mitosporangia, isogamy, anisogamy, oogamy, zygote, sporothallus, alternation of generations

Domain Eukarya
      Supergroup Opisthokonta
            Phylum Zygomycota (Bread Molds)

Related terms: mitosporangiophore, mitosporangia, mitospores, gametangia, zygosporangia, conjugation, homothallic, heterothallic, columella

Domain Eukarya
      Supergroup Opisthokonta
            Phylum Ascomycota (Sac Fungi)
                  Subphylum Pezizomycotina
                        Class Discomycetes
                        Class Plectomycetes
                        Class Pyrenomycetes
                        Class Loculoascomycetes
                  Subphylum Hemiascomycotina (*sensu* Brefeldt 1891)

Related terms: ascoma, ascogonium, antheridium, trichogyne, dikaryon, ascogenous hyphae, hymenium, crozier, karyogamy, endogenous spore development, unitunicate, operculum, operculate asci, ascospores, apical ring apparatus, bitunicate, ascus, apothecial ascoma, perithecial ascoma, stroma, cleistothecial ascoma, pseudothecial ascoma, paraphyses, periphyses, pseudoparaphyses, teleomorph, anamorph, haustoria

Domain Eukarya
      Supergroup Opisthokonta
            Phylum Basidiomycota (Club Fungi)
                  Class Holobasidiomycetes
                        Order Agaricales (Hymenomycetes)
                        Order Polyporales (Hymenomycetes)
                        Order Lycoperdales (Gasteromycetes)
                        Order Sclerodermatales (Gasteromycetes)
                        Order Nidulariales (Gasteromycetes)
                        Order Phallales (Gasteromycetes)
                  Class Phragmobasidiomycetes
                        Order Tremellales (cruciately septate)
                        Order Auriculariales (transversely septate)

Class Teliomycetes
Order Ustilaginales (Smuts)
Order Uredinales (Rusts)

Related terms: monokaryon, clamp connection, dikaryon, basidioma, sterigmata, basidiospore, hymenium, agarics, gills, lamellae, stipe, stalk, cap, pileus, universal veil, volva, partial veil, ring, annulus, peridium, tubes, haustoria, heterocious, alternate host, primary host, spermagonia, spermatia, aecia, transfer spores, aeciospores, uredinial sorus, uredinium, urediniospores (technically the anamorph stage), summer spores, infection stage, telial sorus, teliospores, winter spores, bicellular, abaxial, adaxial

Domain Eukarya
Supergroup Opisthokonta
Phylum Deuteromycota (Fungi Imperfecti)
Class Coelomycete
Class Hyphomycete

Related terms: conidial fungi, conidia, conidiophores, coelomycetes (form conidioma), pycnidium, acervulus, hyphomycetes (no conidioma), amerospore, didymospore, helicospore, staurospore, dictyospore, phragmospore, scolecospore, retrogressive blastic development, basauxic blastic development, thallic development

Domain Eukarya
Supergroup - a composite organism, therefore it belongs to more than one Supergroup!
Phylum Mycophycophyta (Lichens)
Class Lichenes

Related Terms: dual organism (mycobiont+phycobiont), crustose, foliose, fruticose, hymenium, upper cortex, algal symbiont, medulla, lower cortex

Domain Bacteria
Supergroup - are not recognized in this Domain
Phylum Cyanobacteria (Blue-green Algae)

Related Terms: chlorophyll *a*, carotenoids, phycoerythrin, phycocyanin, cyanophycean starch, no flagella, cellulose, pectin, oftentimes sheathed, sometimes calcified, 3.5 billion years, stromatolites, akinetes, heterocysts, hormogonia, mostly fresh water, only asexual reproduction

Domain Eukarya
Supergroup Excavata
Phylum Euglenophyta (Euglenoids)

Related Terms: chlorophyll *a* and *b*, carotenoids, xanthophylls, paramylon granules, 2 apical flagella, no cell wall (pellicle), gullet, ampulla, facultative autotroph, mostly fresh water and pollution tolerant, non-planktonic, phagocytosis, serial endosymbiosis, euglenoid movement, only asexual reproduction

Domain Eukarya
Supergroup Chromalveolata
Phylum Dinophyta (Dinoflagellates)   *Peridium, Gonyaulax, Ceratium*

Related Terms: chlorphylls *a* and *c*, carotenoids, zanthophylls, starch, lipids, 1 girdle, 1 sulcus, cellulose pectin, mostly marine, thecal plates, saxitoxin, paralytic shellfish poisoning, PSP, red tide, filter feeding shellfish, fish kills, asexual reproduction is predominant, halves break and replace, sexual reproduction rarely occurs, zygote, cyst, resting stage

Domain Eukarya
Supergroup Chromalveolata
Phylum Oomycota (Water Molds)

            Order Saprolegniales
            Order Peronosporales

Related Terms: oogonia, antheridia, female gametes, biflagellate zoospore, oogonia with 1 oospore

Domain Eukarya
    Supergroup Chromalveolata
        Phylum Heterokontophyta (Heterokonts)
            Class Bacillariophyceae (Diatoms)
                Order Centrales (Centric Diatoms)
                Order Pennales (Pennate Diatoms)

Related Terms: chlorophyll *a* and *c*, carotenoids, fucoxanthin, chrysolaminarin, 1 tinsel in sperm, silica, pectin, marine and fresh water, frustule, valves, epitheca, hypotheca, girdle, raphe, striae, punctae, asexual reproduction, shrinking diatom syndrome, sexual reproduction, anisogamy, oogamy, fusion of gametes, zygote, centric, pennate, diatomaceous earth, reproductive/evolutionary dichotomy!

Class Chrysophyceae (Golden Algae)   *Synura, Dinobryon*

Related Terms: chlorophyll *a* and *c*, carotenoids, fucoxanthin, chrysolaminarin, 1 tinsel, 1 whiplash, silica, pectin, mostly fresh water, unicellular, colonial, lorica, fishy smelling ketones and aldehydes, asexual reproduction (divide in unicellular form, fragment in multicellular form), sexual reproduction (isogamy)

Class Xanthophyceae (Yellow-green Algae)

Related Terms: chlorophyll *a* and *c* (*c* in low quantities), carotenoids, vaucheriaxanthin, chrysolaminarin, 1 tinsel, 1 whiplash, cellulose, silica in some, mostly fresh water

Class Phaeophyceae (Brown Algae)
        Order Ectocarpales   *Ectocarpus*
        Order Dictyotales   *Dictyota, Padina*
        Order Laminariales   *Postelsia, Alaria, Macrocystis, Laminaria,*
                                        Agarum, Nereocystis
        Order Fucales   *Fucus, Sargassum, Ascophyllum, Hormosira*

Related Terms: chlorophyll *a* and *c*, carotenoids, xanthophylls, laminarin, 2 lateral, 1 whiplash, 1 tinsel, cellulose, pectin, alginic acids, almost all marine, isogeneratae, heterogeneratae, cyclosporae, meiospore, gametothallus, plurilocular gametangium, isogametes, isogamy, zygote, plurilocular sporangium, biflagellate zoospore, mitospore, asexual reproduction, unilocular meiosporangium, meiosis, anisogamous-oogamous, kelps, seaweeds, holdfast, blade, lamina, stipe, intertidal zone, float baldder, alginic acid, meiosporangium, receptacles, conceptacles, paraphyses, antheridium (64 spermatia), oogonium (8 eggs), chemical attractant, alginic acid, fireproofing, laundry starch, lipstick, stabilize foods, etc.

Domain Eukarya
    Supergroup Archaeplastida
        Phylum Rhodophyta (Red Algae)
            Class Bangiophyceae   *Porphyra* (Nori)
            Class Florideophyceae
                Order Ceramiales   *Polysiphonia*
                Order Corallinales   *Corallina*
                Order Gelidiales   *Chondrus* (Irish Moss)
                Order Palmariales   *Palmaria* (Dulse), *Rhodymenia*

Related Terms: chlorophyll *a* (*d*), carotenoids, phycoerythrin, phycocyanin, floridean starch, no flagella, cellulose, pectin, agar, carrageenin, sometimes calcified, mostly marine, tetraspores, male gametophyte, female gametophyte, spermatangium, spermatia, trichogyne, carpogonium, pericarp, carposporothallus, carpospore, tetrasporothallus, tetrasporangium, holdfast, unilocular sporangia, many spermatia per spermatangium, many carpospores per carposporangium versus 1, isomorphic alternation of generations, heteromorphic alternation of generations

Domain Eukarya
    Supergroup Archaeplastida
        Phylum Chlorophyta (Green Algae)
          Class Micromonadophyceae
          Class Chlorophyceae

Order Volvocales	*Chlamydomonas, Pandorina, Volvox, Eudorina*
Order Chlorococcales	*Chlorella, Pediastrum, Hydrodictyon, Scenedesmus*
Order Chaetophorales	*Fritschiella*

          Class Ulvophyceae

Order Ulvales	*Ulva, Enteromorpha*
Order Cladophorales	*Cladophora*
Order Dasycladales	*Acetabularia*
Order Caulerpales	*Caulerpa, Halimeda, Codium*

          Class Charophyceae
            Order Zygnematales (Conjugatophyceae)

Family Zygnemataceae	*Zygnema, Spirogyra*
Family Desmidiaceae	*Cosmarium, Closterium, Staurastrum, Micrasterias*
Order Coleochaetales	*Coleochaete*
Order Charales	*Chara, Nitella*

Related Terms: chlorophylls *a* and *b*, carotenoids, starch, 2-4+ whiplash flagella, cellulose, pectin, sometimes calcified, mostly fresh water, cellular coenocytic, siphonous (multicellular), asexual, sexual reproduction, volvocine series, coenobial, oogamous sexual reproduction, zygote, asexual daughter colony, non-flagellate unicells and colonies, heterotrichous, rhizoid, sporothallus, meiospores, gametothalli, isogametes, quadriflagellate zygote, giant diploid cells with multiple roles (zygote, sporothallus, meiosporangium, gametothallus, gametangium), cytoplasmic streaming, rhizoids, utricles, conjugation, conjugation tube, bilateral symmetry, zygote retention in gametothallus, jacket cells, precursors of land plants

Domain Eukarya
    Supergroup Archaeplastida
        Phylum Marchantiophyta (Liverworts)
          Class Marchantiopsida (Complex-thalloid)
            Subclass Marchantiidae

Order Marchantiales	*Conocephallum, Marchantia, Riccia*

          Class Jungermanniopsida (Leafy & Simple-thalloid)
            Subclass Pelliidae (Simple-thalloid)
            Subclass Metzgeriidae (Simple-thalloid)
            Subclass Jungermanniidae (Leafy)

Order Jungermanniales	*Plagiochila*

        Phylum Anthocerotophyta (Hornworts)   *Anthoceros*
        Phylum Bryophyta (Mosses)

Class Bryopsida (True Mosses)	*Polytrichum, Mnium*
Class Sphagnopsida (Peat Mosses)	*Sphagnum*
Class Andreaeopsida (Granite Mosses aka Lantern Mosses)	

Related Terms: gametophyte, sporophyte, protonema, venter, archegonium, antheridium, neck, zygote, gemmae, gemmae cup, rhizoids, columella, archegoniophore, antheridiophore, seta, capsule, paraphyses, operculum, calyptra, biflagellate sperm, peristomal teeth, foot suspensor, hydroids

Domain Eukarya
    Supergroup Archaeplastida
        Phylum Rhyniophyta (Rhyniophytes)
          Class Rhyniopsida

Order Rhyniales	*Rhynia, Cooksonia*

        Phylum Zosterophyllophyta (Zosterophylls)

Class Zosterophyllopsida
    Order Zosterophyllales
Phylum Trimerophyta
    Class Trimeropsida (Trimerophytes)
        Order Trimerophytales
Related Terms: extinct phyla known only from fossils, dichotomously branched stems, leafless, rootless, homosporous, cauline sporangia, hydroid-like conducting tissues, tracheids

Domain Eukarya
    Supergroup Archaeplastida
        Phylum Lycopodiophyta (Club Mosses)
            Order Protolepidodendrales   *Asteroxylon*
            Order Lepidodendrales   *Lepidodendron, Sigillaria*
            Order Lycopodiales
                Family Lycopodiaceae   *Lycopodium*
            Order Selaginellales
                Family Selaginellaceae   *Selaginella*
            Order Isoetales
                Family Isoetaceae   *Isoetes*
Related Terms: micropyll, sporophyll, strobilus, strobilophore, homosporous, tetrad of spores, bisexual gametophyte, endophytic symbiotic fungi, rhizoids, ligule, heterospory, exospory, endospory, microsporangia, microsporophylls, megaspores, microspores, megasporangia, megasporophylls, trabeculae, velum, archegonia, antheridia, zygote, foot, suspensor, biflagellate sperm, sterile sporophyll, hollow root, rhizophore, cambial layer, 4 megaspores versus extreme heterospory, parichnos scars

Domain Eukarya
    Supergroup Archaeplastida
        Phylum Monilophyta (Ferns and Horsetails)
            Class Psilotopsida
                Order Psilotales (Whisk Fern)   *Psilotum*
                Order Ophioglossales (Grape Fern)   *Botrichium, Ophioglossum*
Related Terms: dichotomous branching, terminal sporangia, sterile branches, lateral sporangia, overtopping, protostele, sporangia, synangia, synangium, eusporangiate
            Class Equisetopsida (Horsetails)
                Order Equisetales   *Equisetum*
                Order Sphenophyllales   *Sphenophyllum*
                Order Calamitales   *Calamites, Annularia*
Related Terms: ribs of sclerenchyma, pith canal, vallecular canals, carinal canals, strobilus, sporangiophore, sporangium, appendicular tip, homosporous, archegonium, antherdium, gametophyte, sporophyte, peltate sporangiophore, fused microphylls, jointed stem, unique lateral branch growth, organ genus
            Class Polypodiopsida (Leptosporangiate Ferns)
                Order Filicales
                    Family Osmundaceae
                Order Marsileales
                Order Salviniales
Related Terms: macrophylls, megaphylls, leptosporangiate, homosporous, heterosporous, exospory, endospory, non-circinnate vernation, circinnate vernation, bisexual gametophyte, fertile pinnule, synangium, fertile/sterile pinnae, rachilla, sorus, annulus, sori, false indusium, reniform indusium, peltate indusium, sporocarp, megasporocarp, microsporocarp, sporophore
            Class Marattiopsida (Marattoid Ferns)
                Order Marattiales
Related Terms: megaphylly, eusporangiate, homospory, endospory, circinnate vernation, bisexual gametophyte, synangial sorus, compound fronds

Domain Eukarya
    Supergroup Archaeplastida
        Phylum Progymnospermophyta (Progymnosperms)   *Archaeopteris, Callyxylon*
        Related Terms: secondary xylem tracheids with circular bordered pits, spores in naked sporangia, organ genera, macrophyllous leaves

Domain Eukarya
    Supergroup Archaeplastida
        Phylum Pteridomspermophyta (Seed Ferns)   *Archaesperma, Caytonia, Medulossa*
        Related Terms: Spermatophyte (macrophyllous leaves with seeds), seed bearing leaves, pollen sac bearing organ, cupule bearing organ, Paleozoic & Mesozoic seed ferns

Domain Eukarya
    Supergroup Archaeplastida
        Phylum Cycadophyta (Cycads)
            Class Cycadopsida
                Order Cycadales
                    Family Cycadaceae
                    Family Stangeriaceae
                    Family Zamiaceae
        Related Terms: Spermatophyte, dioecious, manoxylic (wood high in parenchyma) not pycnoxylic (wood rarely has parenchyma), stems with girdling leaf traces, pollen sac, pollen cone scale, prothallial, antheridial, generative, tube, sterile stalk, fertile body, multiflagellate sperm, microsporangium, micropyle, pollination, pollen droplet, pollen chamber, ovule integument, nucellus, megasporocyte, microsporocyte, archegonial chamber, chalazal end, free-nuclear stage, proembryo

Domain Eukarya
    Supergroup Archaeplastida
        Phylum Ginkgophyta (Maidenhair Trees)
            Order Ginkgoales
                Family Ginkgoaceae   *Ginkgo biloba*
        Related Terms: Spermatophyte, dioecious, long shoots, spur shoots, dichotomously veined, catkin-like pollen cones, paired ovules on peduncle, fleshy seeds, Maidenhair tree, distribution restricted, cultivated, wind pollination, strong apical dominance, auxin, differential anatomy (long versus short shoot) pollen dispersal in four-celled state, female gametophyte development as in the cycads, fleshy, stony, papery layers, butyric acid, satellite chromosomes (female), salnuts

Domain Eukarya
    Supergroup Archaeplastida
        Phylum Gnetophyta (Gnetopyhtes)
            Order Gnetales   *Gnetum*
            Order Ephedrales   *Ephedra*
            Order Welwitschiales   *Welwitschia*
        Related Terms: compound microstrobilus, compound megastrobilus, vessel elements, non-flagellated sperm, fleshy seeds, leaves dicot-like

Domain Eukarya
    Supergroup Archaeplastida
        Phylum Pinophyta (Conifers)
            Order Cordaitales   *Cordaites*
            Order Lebachiales
            Order Pinales

Family Taxaceae   *Taxus, Torreya, Austrotaxus, Paleotaxus*
Family Podocarpaceae
Family Cephalotaxaceae
Family Cupressaceae   *Juniperus, Thuja, Taxodium, Sequoia,*
                                               Sequoiadendron, Metasequoia
Family Araucariaceae   *Wollemia*
Family Pinaceae   *Pinus, Abies, Larix, Picea Tsuga,*
                                               Pseudotsuga

Related Terms: compound ovuliferous cone, simple polleniferous cone, hard pines, soft pines, monoecious, pollination, 4-celled gametophyte, air bladder, megasporocyte, megagametophyte development, microgametophyte development, primary bract, ovuliferous scale, simple polyembryony, cleavage polyembryony, sterigmata, deciduous, quadrangular, petiolate, sessile, fleshy fused scales, dimorphic leaves, knees, General Sherman, General Grant, aril, taxol

Domain Eukarya
   Supergroup Archaeplastida
      Phylum Magnoliophyta (Flowering Plants)
         Class Magnoliopsida (Dicotyledonae)
         Class Liliopsida (Monocotyledonae)

Related Terms: anthophytes, strobilus-like reproductive structures, vessels, placentation, ovary position, axile, parietal, free central, hypogynous, perigynous, epigynous, egg, synergids, polar, antipodal, globular, heart, torpedo, solitary, spike, corymb, raceme, panicle, fruits, double fertilization, flower, co-evolution, pollination biology, sepals, petals, tepals, calyx, corolla, corona, androecium, gynoecium, stamen, anther sac, filament, pistil, carpel, stigma, style, ovary, ovules, double fertilization

Preface

The laboratory component of General Botany provides you the opportunity to view interrelationships between and among structures, to handle live or preserved material, to become familiar with the many terms we use throughout the course, and to learn how to use a microscope properly. Each of you will have your own microscope every week, no exceptions. This laboratory is fundamental, yet integral to your understanding of General Botany. The images in your manual are intended to serve as a guide while you view permanent or prepared slides. These must be viewed by each of you independently. At no time will questions be answered re where is a particular structure, etc., unless the slide is on the stage of your microscope and in focus.

The content of the laboratory is rich, as is the terminology. You must come to lab prepared. You must come to lab knowing what the various terms you are about to deal with mean. There is no such thing as finishing early that simply isn't possible.

In some laboratory exercises you will be asked to identify structures of an organism. For example, Examine slide 9 labeled *Rhizopus* sporangia w.m. and identify the **mitosporangia, mitospores, columella, mitosporangiophore**, and **zygotes**. In all likelihood you will only be able to see **mitosporangia, mitospores, columella,** and **mitosporangiophores.** If zygotes are absent in your slide you note that the population of hyphae you are examining are only reproducing asexually. These questions are written in this manner to further fortify your understanding of the organisms in question and not to trick you. Thinking about what you are viewing is not an option but a necessity!

The phylogeny we have adopted in this course is a composite. No single phylogeny best reflects our collective understanding of all the organisms included in this course so we have created one that reflects modern thought and is based on both morphological and molecular data. None is any more correct or incorrect than is any other, but this is the one that we will use, and the one we deem as most acceptable.

Rest assured, much still needs to be learned about the evolution of many of the groups we will study. Regardless, the course does provide you a general overview of the evolutionary biology of these various groups. This is your starting point, it is not the endpoint!

Classification

We have historically classified objects because it adds order to our lives by facilitating access to these objects, by providing an understanding of them, or by allowing for the easy dissemination of information about them. Basic classification schemes we use today can be found on your hard drive, in the hardware store, the library, a lumberyard, auto wreckers, or your closet. Check out your wardrobe, socks in this drawer, underwear in another, and T-shirts in yet another. One of the major differences between the ways in which you or a respective business may classify their objects is that there is no right or wrong way to construct the groupings – whatever works best for you or the business is deemed acceptable.

Unfortunately the same does not hold true for the biologist. Faced with the daunting task of grouping and naming the world's nearly 2 million species, and in fact the real number may be 14-30 million species, the biologist must adhere to strict naming rules as defined by one of the following: International Code of Nomenclature for algae, fungi, and plants (ICN which replaced the International Code of Botanical Nomenclature – ICBN in July of 2011), the International Code of Nomenclature for Cultivated Plants (ICNCP); the International Code of Zoological Nomenclature (ICZN); International Code of Nomenclature of Bacteria (ICNB); International Committee on Taxonomy of Viruses (ICTV); and the International Code of Phytosociological Nomenclature (ICPN). The work must also be peer reviewed and published in a source acceptable to the scientific community before the newly created group has the potential for having scientific merit. Encompassed within these millions of organisms is a phenomenal amount of intra-group and inter-group variation.

Taxonomy (from the Greek *taxis* for arrangement and *nomia* for method) is that part of the biological sciences that deals with the identification, naming, and arrangement of organisms in a classification scheme. Written taxonomic history can be traced to Theophrastus of Eresos, the intellectual grandfather of modern botany who classified plants as herbs, shrubs, or trees as early as 300 BC. One of the defining periods in taxonomic history occurred during the 18th century and is specifically associated with the work of Carl von Linne (1707-1778), a Swedish botanist known more commonly as, Carolus Linnaeus (Latin was the language of choice among European

015904

HERBARIUM OF SLIPPERY ROCK
UNIVERSITY OF PENNSYLVANIA
(SLRO)

NAME: *Nymphaea odorata*

COMMON NAME: Fraggrant Water Lily

FAMILY: Nymphaeaceae
        Water Lily Family
COLLECTOR(S): Jennifer Evans

DATE: 05-31-98

COLLECTION NO.: 8

LOCATION: Pa, Butler Co.
          Lake Arthur

REMARKS:

scientists so writings and oftentimes names were Latinized) who championed the system of biological nomenclature in common use today. In truth, the binomial system of classification was first proposed by Gaspard (Casper) Bauhin (1560-1624).

Prior to Linnaeus' time there were no commonly accepted standards for naming. As such, organisms tended to have long, cumbersome, descriptive names. For example, botanists referred to the common wild briar rose as *Rosa sylvestris alba cum rubore, folio glabro* which from Latin roughly translates to pinkish white woodland rose with hairless leaves. The Linnaean system of classification brought order to a chaotic science and is founded on the use of binomial nomenclature wherein each of the basic units of classification is given two names (thus a binomial), a genus name and a species name, with the concurrent requirement that these form a pair and are stated together.

Linnaeus referred to the binomial as the *nomen trivial* (trivial name). Today, the binomial, or more specifically, the Latin binomial, constitutes what is referred to as the scientific or specific name. For example, *Quercus alba* is the white oak. Note that the scientific name is italicized and that the first part of the binomial, the genus name is capitalized, whereas the second part, referred to as the specific epithet or species is not capitalized. The genus name is a noun and singular and the species name is usually an adjective but may be a noun. Both are Latinized.

The Linnaean system of classification is viewed as being hierarchical in that each organism is placed within the lowest most group, the species, and closely related species occur in a larger group, the genus (plural genera). Related genera are grouped into a family, families into an order, orders into a class, classes into a phylum (formerly division in the botanical literature), and phyla (plural of phylum) into a Kingdom. Groups of all sizes, from species to Kingdom, are called taxa (singular taxon). The taxonomist is given the latitude to define groups how they choose and in doing so no set number of diagnostic characteristics, no special types of characters, and no special number of characters are involved in the designation of any group of organisms at a particular rank.

All classification schemes have a morphological bias out of necessity. It is possible that during the course of formulating groups that morphological, physiological, ecological, molecular, or chromosome data was used, though when it comes time to actually identify a specimen in hand in the laboratory or field only morphological data is readily available. Further, typically only those characteristics that can be identified with the naked eye or with the use of a hand lens are appropriate for inclusion in the character suite used to define the species. For this reason all classification schemes exhibit a morphological bias.

Although biologists have historically chosen to incorporate aspects of sexual reproductive biology in their definition of the species concept it is prudent not to do so because the strong majority of described species **DO NOT** undergo sexual reproduction. The best workable definition across **all** species boundaries, bacterial, viral, fungal, algal, plant, or animal is:

"a species is the smallest identifiable group of organisms that share more characteristics in common with each other than they do with any other group of organisms".

ALTERNATE LEAF ARRANGEMENT

ANDROECIUM

ANTHER SAC

CALYX

COMPOUND LEAF

COROLLA

DECUSSATE LEAF ARRANGEMENT

DICOT

DIOECIOUS

FERTILE SERIES

FILAMENT

FLORAL SERIES

GENUS

GYNOECIUM

INFLORESCENCE

LATIN BINOMIAL

LEAF BLADE

MONOCOT

MONOECIOUS

OPPOSITE LEAF ARRANGEMENT

OVARY

OVULES

PALMATE VENATION

PARALLEL VENATION

PEDICEL

PEDUNCLE

PERIANTH

PETAL

PETIOLE

PISTIL

RETICULATE VENATION

SEPAL

SIMPLE LEAF

SPECIES

SPECIFIC EPITHET

STAMEN

STERILE SERIES

STIGMA

STYLE

TEPAL

WHORLED LEAF ARRANGEMENT

Label the **sepals, petals, stamens, filaments, anther sacs, pistil, stigma, style, ovary,** and **ovules.**

What type of **placentation** does this flower have? Is this flower **actinomorphic** or **zygomorphic**?

Is this flower **primitive** or **advanced**? What about the following?

Genus: _____   Characteristics

Specific Epithet: _____

Common Name: _____

Family Name: _____

Common Family Name: _____

Genus: _____   Characteristics

Specific Epithet: _____

Common Name: _____

Family Name: _____

Common Family Name: _____

Genus: _____   Characteristics

Specific Epithet: _____

Common Name: _____

Family Name: _____

Common Family Name: _____

Genus: _____   Characteristics

Specific Epithet: _____

Common Name: _____

Family Name: _____

Common Family Name: _____

Genus: _____   Characteristics

Specific Epithet: _____

Common Name: _____

Family Name: _____

Common Family Name: _____

Genus: _____   Characteristics

Specific Epithet: _____

Common Name: _____

Family Name: _____

Common Family Name: _____

Genus: _____   Characteristics

Specific Epithet: _____

Common Name: _____

Family Name: _____

Common Family Name: _____

Genus: _____   Characteristics

Specific Epithet: _____

Common Name: _____

Family Name: _____

Common Family Name: _____

Genus: _____   Characteristics

Specific Epithet: _____

Common Name: _____

Family Name: _____

Common Family Name: _____

Genus: _____   Characteristics

Specific Epithet: _____

Common Name: _____

Family Name: _____

Common Family Name: _____

Genus: _____   Characteristics

Specific Epithet: _____

Common Name: _____

Family Name: _____

Common Family Name: _____

Genus: _____   Characteristics

Specific Epithet: _____

Common Name: _____

Family Name: _____

Common Family Name: _____

Genus: _____   Characteristics

Specific Epithet: _____

Common Name: _____

Family Name: _____

Common Family Name: _____

Genus: _____   Characteristics

Specific Epithet: _____

Common Name: _____

Family Name: _____

Common Family Name: _____

Genus: _____   Characteristics

Specific Epithet: _____

Common Name: _____

Family Name: _____

Common Family Name: _____

Genus: _____   Characteristics

Specific Epithet: _____

Common Name: _____

Family Name: _____

Common Family Name: _____

Genus: _____   Characteristics

Specific Epithet: _____

Common Name: _____

Family Name: _____

Common Family Name: _____

Genus: _____   Characteristics

Specific Epithet: _____

Common Name: _____

Family Name: _____

Common Family Name: _____

Genus: _____   Characteristics

Specific Epithet: _____

Common Name: _____

Family Name: _____

Common Family Name: _____

Genus: _____   Characteristics

Specific Epithet: _____

Common Name: _____

Family Name: _____

Common Family Name: _____

Genus: _____   Characteristics

Specific Epithet: _____

Common Name: _____

Family Name: _____

Common Family Name: _____

Genus: _____   Characteristics

Specific Epithet: _____

Common Name: _____

Family Name: _____

Common Family Name: _____

Genus: _____   Characteristics

Specific Epithet: _____

Common Name: _____

Family Name: _____

Common Family Name: _____

Genus: _____   Characteristics

Specific Epithet: _____

Common Name: _____

Family Name: _____

Common Family Name: _____

Genus: _____   Characteristics

Specific Epithet: _____

Common Name: _____

Family Name: _____

Common Family Name: _____

Genus: _____   Characteristics

Specific Epithet: _____

Common Name: _____

Family Name: _____

Common Family Name: _____

Genus: _____   Characteristics

Specific Epithet: _____

Common Name: _____

Family Name: _____

Common Family Name: _____

Genus: _____   Characteristics

Specific Epithet: _____

Common Name: _____

Family Name: _____

Common Family Name: _____

Genus: _____   Characteristics

Specific Epithet: _____

Common Name: _____

Family Name: _____

Common Family Name: _____

Genus: _____   Characteristics

Specific Epithet: _____

Common Name: _____

Family Name: _____

Common Family Name: _____

Genus: _____   Characteristics

Specific Epithet: _____

Common Name: _____

Family Name: _____

Common Family Name: _____

Genus: _____   Characteristics

Specific Epithet: _____

Common Name: _____

Family Name: _____

Common Family Name: _____

Genus: _____   Characteristics

Specific Epithet: _____

Common Name: _____

Family Name: _____

Common Family Name: _____

Genus: _____   Characteristics

Specific Epithet: _____

Common Name: _____

Family Name: _____

Common Family Name: _____

Genus: _____   Characteristics

Specific Epithet: _____

Common Name: _____

Family Name: _____

Common Family Name: _____

Genus: _____   Characteristics

Specific Epithet: _____

Common Name: _____

Family Name: _____

Common Family Name: _____

Genus: _____   Characteristics

Specific Epithet: _____

Common Name: _____

Family Name: _____

Common Family Name: _____

Genus: _____   Characteristics

Specific Epithet: _____

Common Name: _____

Family Name: _____

Common Family Name: _____

Genus: _____   Characteristics

Specific Epithet: _____

Common Name: _____

Family Name: _____

Common Family Name: _____

Genus: _____   Characteristics

Specific Epithet: _____

Common Name: _____

Family Name: _____

Common Family Name: _____

Readings

Cronquist, A. 1968. The evolution and classification of flowering plants. Boston: Houghton Mifflin.

Linnaeus, C. 1729. *Praeludia Sponsalia Plantarum*. Uppsala. Reprinted 1746, as Sponsalia Plantarum. Stockholm.

Linnaeus, C. 1735. *Systema naturae, sive regna tria naturae systematice proposita per classes, ordines, genera, & species*. Leiden: Haak. pp. 1-12.

Linnaeus, C. 1737a. *Critica Botanica*. Leiden. English ed., 1938. Translated by A. Hort. London: Ray Society.

Linnaeus, C. 1737b. *Genera Plantarum*. Leiden

Linnaeus, C. 1751. *Philosophia Botanica*. Stockholm.

Linnaeus, C. 1753. *Species Plantarum*. 2 vols. Stockholm.

Linnaeus, C. 1754. *Genera Plantarum*. Stockholm.

Linnaeus, C. 1787. The Families of Plants, with their Natural Characters, according to the number, figure, situation, and proportion of all the parts of fructification. Translated anonymously. Lichfield, England: Lichfield Botanical Society.

Mayr, E. 1942. Systematics and the origin of species from the viewpoint of a zoologist. New York: Columbia University Press.

Mayr, E. 1957. Species concepts and definitions. *In:* E. Mayr, (ed.), The Species Problem, pp. 1-22. Washington, D.C.: American Association for the Advancement of Science.

Ruse, M. 1969. Definitions of species in biology. The British Journal for the Philosophy of Science 20: 97-119.

Taxonomy, What's in a Name?

Taxonomy, a term coined by the Swiss botanist Candolle (1813), initially referred to the endeavor of plant classification, that is, the use of morphological characters to order organisms into groups based on their similarities and differences. Today, taxonomy is a more generally used term applied to the act of classifying any group of organisms. Once organisms are placed into a group, that is classified, a taxonomist must apply a unique name to that group following a strict set of globally accepted rules. The naming of groupings of like organisms and the rules that govern the assignment of names is called nomenclature.

"It has been already suggested, and forcibly enough, that plant taxonomy was not invented in any school, or by any philosopher; that it is everywhere as old as language; that no plant name is the name of an individual plant, but is always the name of some group of individuals, and that all grouping is classifying." Edward Lee Greene in Landmarks of Botanical History (1909), page 106.

A Photographic & Ethnobotanical Guide to Plants of Eastern North America

by
Jerry G. Chmielewski

Insofar as Candolle may be credited with coining the term taxonomy, the beginnings of taxonomy must realistically predate recorded history. Further, it could be argued that taxonomy is the world's oldest profession as the earliest of men must have been practical taxonomists out of necessity. As a gatherer society our early ancestors had to be able to communicate with each other regarding the difference between edible and inedible plant matter. Similarly they would have needed to learn about and communicate the difference between for example poisonous and non-poisonous snakes, or the differences among edible, poisonous, and hallucinogenic mushrooms. Clearly their survival was dependent upon their ability to distinguish between or among groups of organisms and concurrently effectively communicate these differences among their social group. Those who failed to learn this distinction in all likelihood also failed to become our ancestors.

Today taxonomy may be subdivided into three branches, alpha or descriptive taxonomy, beta or systemizing taxonomy, and gamma or evolutionary taxonomy. Alpha taxonomy is the level of taxonomy that is concerned with the characterization and naming of species. Many would argue that for all intents and purposes alpha taxonomy is for the most part a piece of the past, though if we consider that only 2 million of the possible 14-30 million species on Earth have been named the potential is staggering. Beta taxonomy is that level of taxonomy concerned with the arrangement of species into a natural system of higher and lower taxa. Relationships between and among species are worked out more carefully and emphasis is placed on sound classification. In truth, Beta taxonomy involves a re-evaluation of the efforts of the alpha taxonomist and entails much of the taxonomy being done today. Gamma taxonomy, also known as evolutionary taxonomy is the level of taxonomy dealing with various biological aspects of taxa including intra-specific studies, speciation, and evolutionary rates and trends and like Beta taxonomy entails much of the taxonomy being done today.

The layperson may argue that the binomial system of nomenclature championed by Linnaeus is too difficult to comprehend because of its Latin roots and as such would prefer to use a system that employs common names instead. Several reasons can be given as to why this approach is impractical: 1. common names are not universal and can have different meanings in different languages; 2. most of the world's organisms do not have common names; 3. many unrelated organisms have been given the same common name; 4. because no rules were established historically for the assignment of common names they have been applied indiscriminately to genera and species; and 5. a single species may be known by more than one common name in the same or different localities.

Although identification may be deemed to be a basic activity it is the primary objective of taxonomy and involves both classification and nomenclature. In its simplest form identification is the determination of similarities and differences between or among objects – they are the same or they are different. The four traditional methods employed to identify unknown specimens include expert identification, recognition, comparison, and the use of keys or similar devices.

Using an expert to identify an unknown specimen would be ideal for many reasons, but realistically, one is typically not on hand for consultation, nor generally concerned with the queries of a layperson. Recognition as a means of identification presupposes past experience and a knowledge base that a neophyte to the world of identification simply would not have. The likelihood of success using the comparison method is very much dependent upon the suitability of the materials used for comparative purposes. Realistically these materials commonly do not extend beyond local field guides or picture books of the more common species and as such are limited in their usefulness. Thus, for the fledgling naturalist the most readily available, reliable, and useable method of specimen identification is through the use of taxonomic keys or similar devices. In the most traditional sense taxonomic keys consist of successive pairs of contrasting statements relating to morphological features of the organism(s) under consideration and culminate in identification. The usage of contrasting pairs of statements, or couplets, in the identification of organisms dates back to Jean de Lamarck in 1778 and has since then been the dominant format employed.

Although the aforementioned key is dichotomous in design and that which is most commonly available and utilized, provision does exist for the possibility of multiple equal statements to be incorporated within a key. Rarely though is the latter implemented. Using a key can be likened to traveling a roadway from point A to B while following directions. Whenever you get to an exit or an intersection you need to make the correct choice to ultimately arrive at the destination you seek. Unlike roadways in which many different routes may be selected to arrive at a certain destination, only a single pathway through a dichotomous key will get you to a specific end point.

Dichotomous keys may be either indented or bracketed. Botanists typically use the indented type whereas zoologists prefer the bracketed type.

Indented Key

    1. Fruit a pome; ovary inferior 2
        2. Petals absent 3
            3. Sepals usually 4; involucre absent .... Plant 1
            3. Sepals usually 5; involucre present .... Plant 2
        2. Petals present .... Plant 3
    1. Fruit a hesperidium; ovary superior 4
        4. Flowers regular; spurs 5 .... Plant 4
        4. Flowers irregular; spur 1 .... Plant 5

To use the indented key read both statements with the lowest number (1) and select that which is most correct. Assuming you have chosen the first of these read both options numbered 2. Assuming the first of these is most correct read both options numbered 3. Select the most appropriate of these and you have identified the specimen to genus.

Bracketed Key

    1. Fruit a pome; ovary inferior 2
    1. Fruit a hesperidium; ovary superior 4
        2. Petals absent 3
        2. Petals present .... Plant 3
    3. Sepals usually 4; involucre absent .... Plant 1
    3. Sepals usually 5; involucre present .... Plant 2
        4. Flowers regular; spurs 5 .... Plant 4
        4. Flowers irregular; spur 1 .... Plant 5

To use the bracketed key read both statements with the lowest number (1) and select that which is most correct. Assuming you have chosen the first of these next go to the statements numbered 2 – read both options. Assuming the first of these is most correct go to the statements numbered 3 and read both options. Select the most appropriate of these and you have identified the specimen to genus.

One of the primary drawbacks in the use of dichotomous keys by the non-professional rests in the typically technical terminology incorporated within the couplets. As you read through either the indented or bracketed key you should have noticed that several technical terms are included. For example, do you know what is meant by achene, follicle, spur, petal, sepal, involucre or flowers being regular versus not? Not knowing the terminology specific to the type of organism you are trying to identify is the major drawback to the use of dichotomous keys. Phytography is that part of taxonomy which deals with the description of plant parts. Although its intent is to portray accuracy and completeness of description in the fewest number of words possible it is a point of frustration for the neophyte user. This limitation can however be easily remedied through the use of less technical, though equally accurate descriptive terminology. In reality, the trend in newer keys is to use terminology directed toward usage by the non-professional as opposed to the professional biologist. The second primary restriction of dichotomous keys is that by their very structure a single point of entry is mandated. That is, the user must begin at the first couplet in the key and sequentially progress through subsequent couplets until the respective specimen is identified. Whether intended or not, the single point of entry implies a hierarchy in the characters used to distinguish among specimens. Greater latitude in this regard is however provided through polyclonal, or synoptical, keys. These multi-entry, or any-order, keys have historically used cards stacked in any arrangement, one on top of the other, with holes or edges punched such that cards with the desired taxa to be retained are eliminated until the card with the desired taxa listed on it is the only card remaining. Today though, the personal computer offers the possibility for automated identification using multi-entry keys as well as key construction *per se*. Further, the personal computer also permits microcomputer-assisted telephone identification of plants and a new mobile app can be used to identify plants by leaf shape. How far we have come since the days of Linnaeus.

The current exercise uses a personal computer to run a multi-entry key that is complemented by digital images to be used for comparative purposes to identify native, naturalized, or cultivated conifer genera that are included in the flora of North America north of Mexico. No a priori knowledge of conifer genera is necessary to successfully navigate through the key nor is knowledge in the use of, or creation of, a key necessary.

Because our key includes characters that specifically deal with character states of leaves/needles and seed cones both would be preferred to be on hand for material being identified, though not required. Experience has taught us that in all likelihood leaves/needles will more typically be available for identification purposes than would seed cone material. This said, regardless of material on hand, a correct identification would still be likely because the user could toggle among the available digital images of remaining choices and use the comparison method of identification to arrive at a conclusion.

The conifers are the ideal group of organisms to use for such a learning exercise not only because they are common components of the native vegetation as well as being commonly cultivated, but because they maintain their character states throughout the year as opposed to only part of the year such as is found with deciduous trees or herbaceous vegetation. Thus, irrespective of the time of the year or when an instructor chooses to incorporate the exercise they would be able to do so.

Summary

The four basic components to taxonomy include description (describing the individual entities), classification (placing the entities into groups), identification (the process of placing an entity into an existing group), and naming (assigning a name to a newly discovered entity). Few of us prepare formal classification schemes, describe taxa, or assign names to these; however, many of us are users of these resources in that we attempt to identify what are unknown taxa to us.

Glossary

Acute - ending in a point

alpha taxonomy - the level of taxonomy concerned with the characterization and naming of species – also known as descriptive taxonomy

anthesis - the period during which the anthers are releasing pollen

appressed - lying close to or parallel to an organ, as hairs appressed to a leaf or leaves appressed to a stem

aristate - with a bristle terminating a bract

beta taxonomy - the level of taxonomy concerned with the arrangement of species into a natural system of higher and lower taxa – also known as systemizing

binomial nomenclature - a Latinized two part name for an organism, developed by Carolus Linnaeus, consisting of the genus (first part) and species (second part)

Class - a category within the hierarchical system of classification that is composed of a group of similar Orders

classification - the use of morphological characters to order organisms into groups based on their similarities and differences

comparison method of determination - involves comparing an unknown with named specimens through the use of photographs, illustrations, or descriptions

decurrent - extending downward

dentate - margins with rounded or sharp coarse teeth that point outward at right angles to the midvein

descriptive taxonomy - the level of taxonomy concerned with the characterization and naming of species – also known as alpha taxonomy

dichotomous key - an identification tool that uses a series of contrasting statements, known as couplets, to identify unknown organisms

dimorphic - existing in two forms

Division - a category within the old hierarchical system of plant classification that is composed of a group of similar Classes (equivalent to phyla in today's treatment)

Domain - the most inclusive group in a taxonomic hierarchy where cellular and molecular characteristics distinguish groups from one another.

erose - irregularly cut or toothed along the margins

evolutionary taxonomy - the level of taxonomy dealing with various biological aspects of taxa including intra-specific studies, speciation, and evolutionary rates and trends

expert determination - use of a recognized professional in the discipline to identify an unknown object with which they have familiarity

exserted - projecting out or beyond; often referring to stamens or styles which project beyond the perianth

Family - a category within the hierarchical system of classification that is composed of a group of similar Genera

gamma taxonomy - the level of taxonomy dealing with various biological aspects of taxa including intra-specific studies, speciation, and evolutionary rates and trends – also known as evolutionary taxonomy

Genera - plural of genus

Genus - in all likelihood the oldest concept of taxa developed by mankind consisting of species

glabrous - lacking hairs, smooth

glaucous - gray, grayish-green or bluish with a thin coat of fine removable waxy particles

imbricate - shingled; overlapping, as shingles on a roof

keeled - having a sharp or conspicuous longitudinal ridge, like the bottom of a boat

Kingdom - a category within the hierarchical system of classification that is composed of a group of similar Phyla or Divisions

lanceolate - shaped like a lance-head, much longer than wide and widest below the middle

mucronate - sharply and stiffly pointed

nomenclature - the naming of groupings of like organisms and the rules that govern the assignment of names

obtuse - margins straight to convex, forming a terminal angle greater than 90°; blunt, rounded

Order - a category within the hierarchical system of classification that is composed of a group of similar Families

ovate - egg-shaped

petiolate - borne on a stalk

Phyla - plural of Phylum

Phylum - a category within the hierarchical system of classification that is composed of a group of similar classes (equivalent to Division in the old hierarchical system of plant classification)

phytography - that part of taxonomy that deals with descriptions of plants and their various organs

pubescent - hairy

recognition - requires that you as an individual be sufficiently experienced to make the identification

reflexed - bent backward

retuse - indented, notched slightly

sessile - without a stalk or a petiole, attached directly by the base

species - the smallest identifiable group of organisms that share more characteristics in common with each other than they do with any other group of organisms

spinulose - having little thorns or small spines over the surface

sterigma - short stalk joining leaf to stem, persistent after the leaf/needle falls off

stipitate - with a stipe or stalk

subacute - more or less acute

subulate - awl-shaped; bearing sharp points

Supergroup - a level in the taxonomic hierarchy for Eukaryotic life that is less inclusive than Domain, but more inclusive than Kingdom.

taxon - a specific classification category applied to a particular group of organisms

taxonomist - a scientist specializing in taxonomy

taxonomy - the science of classification concerned with the processes of classification, nomenclature, and identification

umbo - a rounded elevation or protuberance at the end or side of a solid organ

PART I: Using a dichotomous key

Key to Common Conifer Genera

1. Leaves in a definite cluster ………………………………………………..……………….... 2

   2. Leaves (needles) in clusters of 2-5, each cluster surrounded by a membranous sheath … Pine
Pinus
Pinaceae (Pine) Family

   2. Leaves (needles) in clusters on short lateral branches …..………………….... Larch, Tamarack
Larix
Pinaceae (Pine) Family

1. Leaves not in definite clusters …………………………………………..…………………… 3

   3. Leaves alternate or scattered ………………………………………………………………..… 4

      4. Leaves joined at the base to a spreading or appressed sterigma (a raised region) that is basally decurrent (bent backwards) on the stem and persistent after leaf fall
…………………………………………………………………………………………… 5

         5. Leaves sessile on the sterigma, quadrangular in cross section …….....................… Spruce
Picea
Pinaceae (Pine) Family

         5. Leaves petiolate on the sterigma, flat in cross section ……………………..…………….... 6

            6. Leaf tips blunt or retuse (shallow notch), marked with two white lines on the underside ……………………………...………………………………………….... Hemlock
Tsuga
Pinaceae (Pine) Family

            6. Leaf tips acute (pointed), green on both sides, but darker on the upper surface
……………………………………………………………………………………………... Yew
Taxus
Taxaceae (Yew) Family

      4. Leaves attached directly to the branch …………………………………..……………….... 7

         7. Leaves not flat ……………………………………………………..……………… Pond Cypress
Taxodium
Taxodiaceae (Redwood) Family

         7. Leaves flat ……………………………………………………………..…..……………… 8

8. Leaves petiolate (with a leaf stalk) .... Douglas Fir

Pseudotsuga

Pinaceae (Pine) Family

8. Leaves sessile (attached directly to the branch) ...…...............…...................... 9

9. Branchlets persistent, leaves eventually deciduous, leaving a round smooth scar .... Fir

Abies

Pinaceae (Pine) Family

9. Branchlets deciduous with their leaves, branches without leaf scars ...……....... ...…..…... Bald Cypress

Taxodium

Taxodiaceae (Redwood) Family

3. Leaves distinctly opposite or whorled, minute 10

10. Leaves not dimorphic in alternating pairs, leafy twigs not flattened ...……...... Juniper

Juniperus

Cupressaceae (Cedar or Cypress) Family

10. Leaves dimorphic in alternating pairs, the lateral ones folded and often keeled, those of the upper and lower surfaces flat or somewhat convex ....…....................... 11

11. Leafy branches somewhat flattened, half to two-thirds as thick as wide .... Cedar

Chamaecyparis

Cupressaceae (Cedar or Cypress) Family

11. Leafy branches strongly flattened, a fourth as wide or even narrower ....Arborvitae

Thuja

Cupressaceae (Cedar or Cypress) Family

Use the Tree Finder Key provided to identify the conifers to species. Keep in mind that *Taxus* (the Yew) is not considered a tree and is not in the Tree Finder Key.

SPECIMEN A

White pine

Pinus strobus

SPECIMEN B

~~Dawn redwood~~   X

~~Metasequoia glyptostroboides~~

Tsuga canadensis

SPECIMEN C

White pine

Pinus strobus

SPECIMEN D

Yew

Taxus baccata

SPECIMEN E

~~Hemlock~~   X

~~Tsuga canadensis~~

Juniperus thyoides

SPECIMEN F

~~White spruce~~

~~Picea glauca~~

Picea abies

SPECIMEN G

~~Red cedar~~   X

~~Juniperus virginiana~~

Chamaecyparis thyoides

SPECIMEN H

Arborvitae

Thuja occidentails

SPECIMEN I

Blue spruce

Picea pungens

SPECIMEN J

~~Norway spruce~~

Picea ~~abies~~
glauca

PART II: Making a dichotomous key

STEP 1 – decide whether you are going to create an indented or bracketed key
STEP 2 – identify all the groups to be included in the key and prepare a detailed description of each
STEP 3 – select those macroscopic characters that have contrasting character states to distinguish between and among groups

SUGGESTION 1 – use parallel construction and comparative terminology in each couplet (e.g. leaves alternate versus leaves opposite)
SUGGESTION 2 – if possible use two characters per couplet
SUGGESTION 3 – start the couplet with the part of the organism being described (e.g. leaves green not green leaves)

Prepare a dichotomous key to the specimens provided by your instructor.

Group 1 characteristics

Needle-like

Group 2 characteristics

Group 3 characteristics

Group 4 characteristics

Group 5 characteristics

How many couplets are necessary to distinguish among 5 specimens? What is the relationship between the number of specimens to be classified and the number of couplets?

Indented or Bracketed Key (circle that which you are preparing)

Juniperus Virginana - Red cedar

Yew

Picea glauca - white spruce

Thuja Occidentalis - Arborvitae

Tsuga Canadensis - Hemlock

Picea Abies - Norwa

Pinus Resinosa - Red pine

Metasequoia glyptostroboides - Dawn Redwood

Picea Pungens - blue spruce

Pinus Strubus - white pine

☆ know these + pictures ☆

PART III: Using a multi-entry point computer based identification program

The identification program you are about to use is called CID and was designed to identify conifer specimens to genus. The program uses both a multi-entry point computer based identification program as well as the comparison method for identification of unknowns.

How to use the program…

STEP 1 – turn the computer on
STEP 2 – follow the instructions at the logon window
STEP 3 – double click on the *Conifer ID* icon

Once the window opens note the contents of the bottom right hand corner. Five character options are available for selection, *leaf position, leaf shape, leaf association, leaf arrangement*, and *fruit type/position*. Select any of these characters as the point of entry. For the sake of simplicity let's enter the key at leaf position – note that leaf position is highlighted in blue. The box on the upper left hand side of the window asks the question Do the leaves/needles occur on branches and/or branchlets or do they occur on short spur shoots? If any terms within this question are unknown to you move the cursor over the question and a box will open explaining the more technical terms. Assuming that you want clarification of the two options move your cursor over the uppermost question mark and left click the mouse. Note that a picture appears in the box on the upper right of the screen demonstrating an example of the character in question. The name or web address on the lower left hand corner of the picture signifies the source of the picture. A magnifying glass in the lower right hand corner of the picture indicates that this window may be enlarged – move the cursor over the magnifying glass and left click the mouse. Once you are done viewing the magnified view minimize or close the window. Move the cursor over the lowermost of the two question marks and repeat the procedure above. Note that at the bottom of this window is a line indicating that 25 of 25 items remain, that is that 25 different selections exist in the data set.

STEP 4 – select either the first or second (let's select this one as we proceed through the tutorial) option by clicking on the open circle

Note that by selecting one of the character options that a new picture appears in the window to the right, that the arrow beside the *leaf position* character in the bottom right hand window switches from red to green and that the number of items remaining in the data set changes from 25 to a lower number of possibilities.

STEP 5 – select the next character by either using the green arrows in the bottom left hand box or clicking on a character in the box on the bottom right

For the sake of simplicity highlight *leaf shape*.

The window to the upper left now has the question. What is the basic form of the leaf? If you have questions about the question per se or issues with the character states follow the procedures from above.

STEP 6 – select either the first (let's assume this is the correct response in this instance), second, or third option by clicking on the open circle

Note that once the character state is highlighted that the number of items remaining in the data set drops from value to a lesser value and that a picture appears in the window to the right.

STEP 7 – select the next character to be examined (*leaf association*) following the procedure from above

Note that the question 'How do respective leaves occur in association with other leaves?' appears in the box and that of the 11 options only 5 are highlighted

STEP 8 – select the most appropriate response for the material on hand (let's assume the last, 5 per sheathed bundle, is the most correct)

Note that at the bottom of the upper left hand window only 1 of 25 items remains. The window to the upper right includes a picture of Pine type E (*Pinus*) that can be magnified. Because only one item remains in the data set the identification is complete. Does the picture and object in hand match? If the specimen in hand and picture do not match go back to the beginning and see where you went wrong. Note that in this instance we didn't even need to use either the fourth or fifth character to complete the identification.

SPECIMEN A   SPECIMEN B

SPECIMEN C   SPECIMEN D

SPECIMEN E   SPECIMEN F

SPECIMEN G   SPECIMEN H

SPECIMEN I   SPECIMEN J

PART IV: Using the comparison method to identify an unknown

STEP 1 – place the cursor over either the forward or backward key in the upper right hand box and press the left mouse

Note that the picture will change with each click of the mouse as will the appropriate character state in the upper left hand box (it will turn from black to blue). By selecting different characters in the lower right hand box and scrolling through the different pictures you can watch character states change.

STEP 2 – scroll through the photographs until you arrive at that which is most similar to the specimen being identified

STEP 3 – enlarge the image if possible and make your final comparison

SPECIMEN A   SPECIMEN B

SPECIMEN C   SPECIMEN D

SPECIMEN E   SPECIMEN F

SPECIMEN G   SPECIMEN H

SPECIMEN I   SPECIMEN J

Readings

Cronquist, A. 1968. The evolution and classification of flowering plants. Boston: Houghton Mifflin.

Metsger, D.A. 1990. Microcomputer-assisted telephone identification of plants in response to poison control calls. Journal of Toxicology Clinical Toxicology 28: 135-157.

Smithsonian Science. Leafsnap, a new mobile app that identifies plants by leaf shape, is launched by Smithsonian and collaborators. http://smithsonianscience.org/2011/05/new-mobile-app-that-identifies-plants-by-leaf-shape-launched-by-smithsonian-and-columbia-and-maryland-universities/. Accessed January 25, 2012.

Strain, S.R. & Chmielewski, J.G. 2010. A simple computer application for the identification of conifer genera. American Biology Teacher 72: 301-304.

Hart, E.W. 1938. Keys to goldenrod in Canada and Newfoundland. Contributions to Canadian Botany 565: 1-31.

Watson, L, Aiken, S.G., Dallwitz, M.J., Lefkovitch, L.P., & Dube, M. 1986. Canadian grass genera: keys and descriptions in English and French from an automated data bank. Canadian Journal of Botany 64: 53-70.

Watts, M.T. 1991. Tree Finder: A manual for the identification of trees by their leaves. Nature Study Guild Publishers. Rochester, NY, USA

Microscopy

The microscope was invented in the 1590's by Zacharias and Hans Janssen of the Netherlands. The early microscope was merely a simple tube with a series of lenses and a mirror that gathered natural light. These microscopes could magnify structures 3-10 times that of the unaided eye. It was nearly seventy years later that Robert Hooke applied the use of the microscope to biological objects when he described/discovered the cellular structure of organisms. The "light" microscope revolutionized our ideas of organismal structure in that it allowed for the elucidation of fine structure.

Today, the microscope is an instrument that is oftentimes used in the biological sciences. Two main types of microscopes are commonly utilized, namely, the light microscope and the electron microscope. Light microscopy includes: conventional light microscopy, phase contrast microscopy (light is phase shifted to increase contrast), fluorescence microscopy, and confocal microscopy (uses point illumination to reconstruct three dimensional objects). In both fluorescence and confocal microscopy fluorescent stains are used to generate images. These are commonly used to examine the fine structures within cells. Electron microscopy includes scanning electron microscopy (SEM) and transmission electron microscopy (TEM). These techniques use electrons to illustrate the surface architecture of cells and the fine structures within cells, respectively. In this course we will use conventional light microscopy throughout the semester. Electron microscopy, fluorescence microscopy, and confocal microscopy will be utilized in upper division courses.

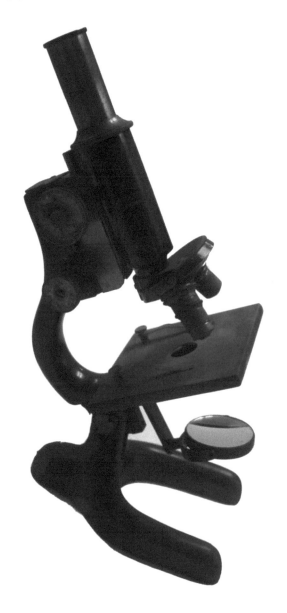

PART I: Introduction to the compound microscope

Much of your time in lab will be devoted to viewing specimens with either your compound microscope or as we will see in the next section the dissecting or stereo microscope. Your lack of familiarity with the various controls that focus light or the specimen *per se* will drastically impede your ability to see what you are expected to see, as well as increase the likelihood of damage to the microscope or slides. None of these outcomes is acceptable. Using a microscope is not as simple or easy as people think. In the past we have found that even upper-year students often have little or no idea how to store, transport, or operate a microscope properly. This is truly disappointing. Today's lab is designed to get you over this hurdle. As a biologist, you will spend many hours using microscopes. It takes a little technical 'know-how' and a lot of care to get the most out of microscopy. Please learn the basics now, and save yourself (and me) a lot of grief.

Microscopy will present you with four major challenges.

1. Bringing the material into focus
2. Selecting the most informative parts of the slide
3. Interpreting and relating to what you see
4. Recording observations for future reference

It is now time to bring a microscope to your bench. Remove a microscope from the cupboard by placing one hand below the base and the other near the top. Remove the microscope from the cupboard while maintaining it in an upright position being careful not to hit adjacent microscopes. Always use two hands to carry your microscope from the cupboard to the bench and vice-versa. Never slide your microscope across the bench top.

Identify and become familiar with the following parts of your microscope

1. **Ocular Lens (eye piece)** – The eye piece magnifies the sample 10× what you see with your unaided eye.

2. **Objective Lens (objective)** – The objective lens magnifies the sample the amount of times indicated on the objective lens. For example, the 4× objective lens magnifies what you see with your unaided eye 4×.

On the compound microscopes you will be using there are generally three objective lenses. These are the 4×, 10×, and 40×.

$$4× \quad = \quad \text{scanning objective}$$

$$10× \quad = \quad \text{low power objective}$$

$$40× \quad = \quad \text{high power objective}$$

You will notice two sets of numbers on the objective lens (10×/0.25 – SEE IMAGE BELOW). The first number (10×) refers to the magnification of the objective lens. The second number (0.25) refers to the Numerical Aperture (NA). The higher the value of the NA of a lens the greater the resolving power of that objective. This means the greater the ability you will have to distinguish between two points. Numerical Aperture is equal to $n \sin \theta$ where n is the index of refraction of the medium in which the lens is working (i.e. 1.0 for air, 1.33 for pure water, and up to 1.56 for oils and θ is the half angle of the maximum cone of light that can enter or exit the lens. The size of the finest detail that is resolved is proportional to λ / NA, where λ is the wavelength of light. This said, a lens with a higher NA provides better detail than one with a comparatively lower NA. Lastly, lenses with a larger NA also collect more light than do those with a comparatively lower NA, thus resulting in a brighter image).

Typical NA Values

4×	0.1
10×	0.25
40×	0.65
100×	1.2

Magnification of the image

The total magnification of a specimen can be determined by multiplying the magnification of the **ocular lens** by the magnification of the **objective lens**.

Total magnification = magnification of **ocular lens** X magnification of **objective lens**

What is the magnification of the sample if you are using a 10× **ocular lens** and a 40× **objective lens**?

3. **Revolving nosepiece (nosepiece).** The objectives attach to the microscope with the nosepiece. By rotating the nosepiece the objectives may be changed. This allows you to increase or decrease the overall magnification.

4. **Stage.** This is the part of the microscope that supports the slide. Our microscope is equipped with spring clamps to hold the slide in place. The stage is mechanical, that is, by moving the slide you change your field of vision. Remember that what you see through the eyepiece is upside down and backwards in relation to what you see on the stage. In the center of the stage is an opening beneath which is an aperture that allows focused light to pass through the slide to the objective lens.

5. **Condenser lens.** The aperture mentioned above is the condenser lens which concentrates light from the light source below and focuses it on the specimen.

6. **Condenser lens diaphragm (iris diaphragm).** The iris diaphragm is used to control the diameter of the beam of light that passes through the condenser lens. For new users this is an oftentimes overlooked adjustment that provides for contrast.

7. **Condenser lens focus knob.** This knob moves the condenser lens up and down to focus the light on the specimen. This knob is often overlooked by students as being useful.

8. **Coarse focus knob.** This knob moves the stage up or down large distances to bring the specimen into rough focus. The coarse focus knob should only be used with the lowest objective in place.

9. **Fine focus knob.** This knob moves the stage up or down small distances and brings the specimen into sharp focus.

10. **Illumination source.** This is the light source of your microscope and is controlled by both the on/off switch and light intensity knob or lever.

PART II: Using the compound microscope.

Your instructor will now lead you through an exercise on how to properly use and store your microscope.

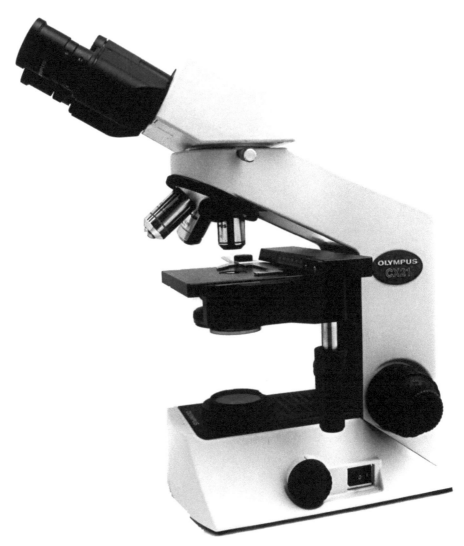

PART III: Set up your compound microscope for Köehler illumination.

Normally this would be performed with the condenser lens focus knob and an illumination source diaphragm, but because your microscope lacks the latter we will need to improvise.

1. Turn on the microscope.

2. Move an objective in place.

3. Look through the eyepieces.

4. Place part of your thumb over the illuminator so that you can see your thumb (in part) and the light.

5. Move the condenser lens focus knob until your thumb is in focus.

6. Your microscope is now set up for Köehler illumination

PART IV: Dissecting or Stereo microscope

1. Examine the dissecting/stereo microscope noting the similarities and differences that exist between it and the compound microscope. How does it compare physically with a compound microscope?

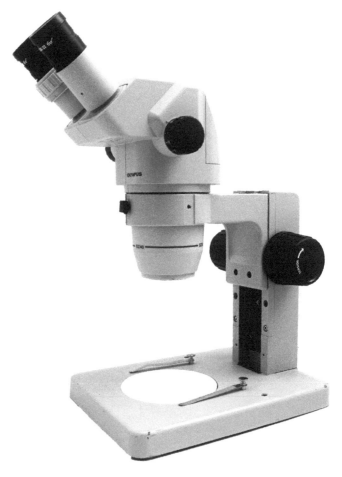

2. Why does the dissecting/stereo microscope lack a fine focus knob?

3. Are the uses of the compound and dissecting/stereo microscopes comparable? Explain.

PART V: Examination of a prepared slide

1. Obtain a prepared slide of silk or hemp fibres.

2. Identify the following: label which includes manufacturer and slide content information, and coverslip. wards Hemp fibers photem w.m.

3. Clean the top of the coverslip and the slide beneath the coverslip with a Kimwipe that has been dipped in alcohol. Note: Do not press down on the coverslip.

4. Examine the silk or hemp fibres using both the compound and dissecting microscopes.

   a. Starting with the 4× objective note the magnification of the sample for each objective.

   b. Remember to use the coarse and fine focus to sharpen the image of your specimen. Note: Do not use the coarse focus with the 40× objective. Note: Use care not to run the slide into the objectives because it will break the slides and scratch the objective lens. Many of these slides are irreplaceable and the objective lens is expensive.

5. Your microscope is parfocal which means that as you switch from one objective lens to another the image should stay in focus or nearly so. Only the fine focus should be used to sharpen the image.

PART VI: Preparation of a wet mount of *Rhizopus*

In this course we will not only use permanent slides that have been purchased from a biological supply house, but also slides that you will prepare yourself from live material – these are known as wet mounts. Permanent slides are a convenient way of re-investigating a sample at a later date. If cared for properly these can be used by students like you for decades. We will use two types of permanent slides, whole mounts and sectioned mounts. In whole mounts the specimen may be stained or cleared so that external structure 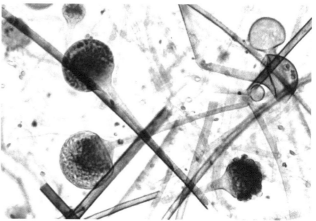 is visible. To generate a sectioned permanent slide a microtome and differential staining are used. The resultant mount allows the investigator to view more of the specimen's internal fine structure than is possible with whole mounts.

Wet mounts (live mounts) will also be used in this course. Today you will make your own slide preparation of the zygomycete *Rhizopus*. Place a drop of distilled water in the center of a clean slide. Use the tip of a dissecting needle or mounting pin to remove a small amount of *Rhizopus* from the culture and place this into the drop of water. As a general rule of thumb less is best when it comes to preparing wet mounts. Gently place a coverslip over the wet mount by first touching it to the drop of water at a 60 degree angle and letting it fall into place. Examine your wet mount using different objectives. Use your mycology skills to identify **spores**, **columella** and **sporangiophores**. This sample is a wet mount, but it is also a whole mount in that it is a sample of the entire organism as opposed to a section of it.

PART VII: Final steps regarding microscope use

Always carry out the following steps before returning your microscope to the cupboard at the end of the laboratory period:

1. Rotate the 4× objective into place.
2. Lower the stage.
3. Clean the objectives with lens paper NOT Kimwipes.
4. Clean the stage with a damp Kimwipe.
5. Unplug and wrap the electrical cord around the base.

Readings

Ford, B.J. 1998. The Earliest Views. Scientific American 278: 50-53.

Smith, G.F. & Tiedt, L.R.. 1991. A rapid, non-destructive osmium tetroxide technique for preparing pollen for scanning electron microscopy. Taxon 40: 195-200.

Domain Eukarya
    Supergroup Amebozoa
        Phylum Myxomycota (Slime Molds)
            Class Myxomycetes (Plasmodial)
            Class Dictyosteliomycetes (Cellular)
            Class Protosteliomycetes (Protostelids)

Related Terms: myxamoebas, phagocytosis, pseudoplasmodium, macrocyst, plasmodium, sporangia (advanced), aethelia, plasmodiocarp (primitive)

Slime Molds

The slime molds (Phylum Myxomycota) are an eclectic group of fungal-like protists that because of a convoluted history have been included in both introductory botany and zoology courses. Three major groups of slime molds are recognized, the plasmodial or acellular slime molds (Class Myxomycetes), the dictyostelids or cellular slime molds (Class Dictyosteliomycetes), and the protostelids (Class Protosteliomycetes). The life history strategies exhibited by members of these Classes are varied though include an assimilative phase and a reproductive phase. All share a structurally similar fruiting body that consists of a cellulosic stalk of one to many sterile cells that support the spore bearing structures. No unanimity of opinion exists with respect to whether the group is monophyletic or polyphyletic. Truth be told, the fungal-like plasmodia of the myxomycetes and the motile slug-like stage or pseudoplasmodium of the dictyosteliomycetes in concert with the plant like fruiting bodies of both has resulted in their classification throughout time as animals, fungi or plants. Molecular phylogenies of rRNA genes show little support for coherence within the slime molds. However, actin and beta-tubulin trees place the myxomycetes and dictyosteliomycetes together. Further, the slime molds collectively and consistently, though not 100 percent of the time, are placed closer to animals and fungi than to green plants in phylogenetic reconstructions. Molecular phylogenies based on protein synthesis elongation factor-1α (EF-1α) support the monophyly of the slime molds. Inasmuch as it is probably more appropriate to include these organisms in a general zoology course historical precedence dictates their inclusion in a general botany course, thus their inclusion here.

Class Myxomycetes (plasmodial or acellular slime molds)

The Myxomycetes are represented by less than 1,000 species. They lack a cell wall and are heterotrophic, feeding on bacteria, spores and decaying organic matter. Cells may be aflagellate and amoeboid-like (myxoamoebae) or flagellate (swarm cells). Many species, such as those in the genus *Physarum*, have a life cycle that is categorized as sporic meiosis (zygote divides mitotically to produce a multicellular sporophyte which then produces spores through meiosis and these divide mitotically to produce a haploid gametophyte) where there is both a free living haploid and diploid generation. The fusion of two haploid cells results in the formation of a free living diploid phase which is a coenocytic plasmodium (multinucleate mass of cytoplasm that lacks internal septation). Once food supplies are exhausted the plasmodium may differentiate into one of three types of sporulating structures (aethelium, plasmodiocarp, or sporangium). Alternatively, the plasmodium may differentiate into a resting structure called a sclerotium.

Physarum a plasmodial slime mold

Physarum polycephalum moves in an amoeboid fashion ingesting solid food particles through phagocytosis as would an amoeba and concurrently absorbs dissolved nutrients. It is no wonder then that this organism is better dealt with in general zoology than here, though the reproductive structures would dictate otherwise. Regardless, as we have adopted a historical approach to this course they are treated here. While examining the petri-dishes with cultures of *Physarum polycephalum* note the fan-shaped network of vein-

like strands. Each of the vein-like strands of plasmodium consists of a hyaline (glassy, transparent, or translucent), semisolid, outer layer (gel state) of protoplasm and a fluid inner portion. Use your dissecting microscopes to examine the rhythmic streaming within the plasmodium as it flows in one direction for a few seconds, slows to a stop, and then reverses the direction of flow. Cytoplasmic streaming, or cyclosis, is facilitated by the contractile protein myxomyosin which behaves like actinomyosin in animal muscles, changing its viscosity when ATP is added. Why does the flow change direction? Why does streaming even occur? The distinctly granular nature of the protoplasm is a consequence of pigment granules, bacterial cells, and other materials ingested by the plasmodium.

1. Examine the culture of *Physarum polycephalum* a plasmodial slime mold. Note the **plasmodium.**

    a. How many cells make up the plasmodium?

    one
    one cell

    b. Is the plasmodium **haploid** or **diploid**?

    2 haploid fuse to
    make a diploid zygote

    c. What type of life cycle does this organism exhibit?

    sporic meiosis

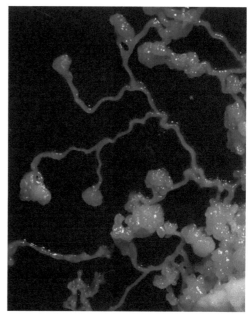

d. Is this a **cellular** or an **acellular slime mold**?

e. To which Phylum does this organism belong?

Myxomycoda

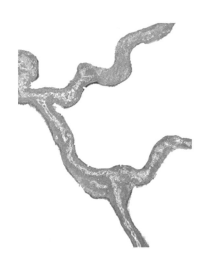

2. Examine dried specimens of the plasmodial slime molds. Identify the **sclerotia**, **plasmodiocarp**, **aethelium**, and **sporangium** if present.

Sclerotia

Plasmodiocarp

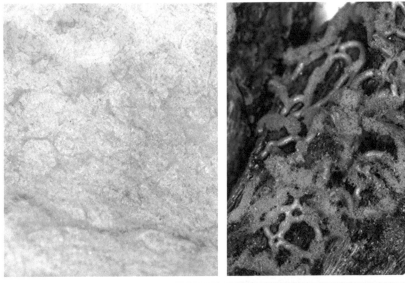

Aethelium

Sporangium

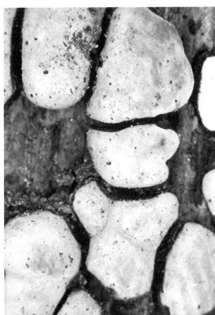

3. Examine slide 2 labeled *Physarum polycephalum*. Explain what you see?

- many nuclei in plasmodium

- no cell walls

- cytoplasmic streaming

- free-living 2n cell

4. Examine slide 1 labeled *Arcyria* capillitium and spores. *Arcyria* has long, coiled threads (capillitium) inside the sporangium – the capillitium gradually straightens out (like a slinky coming out of a can) flipping the spores off into the air. Identify those structures with which you should be familiar. To which Class does *Arcyria* belong? To what specific structure in the life cycle of a Myxomycete is the capillitium associated? Is this the sexual or asexual phase?

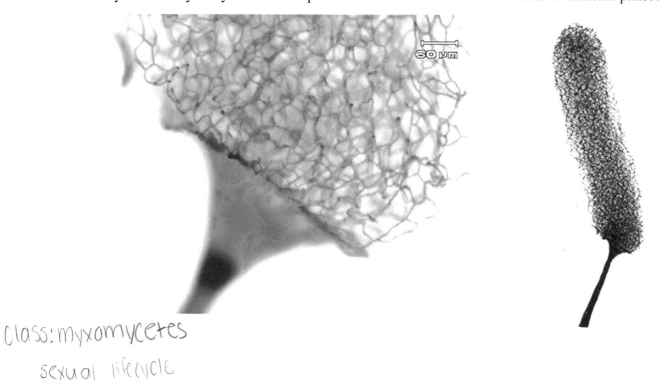

Class: myxomycetes

sexual lifecycle

5. Illustrate the life cycle of *Arcyria* below.

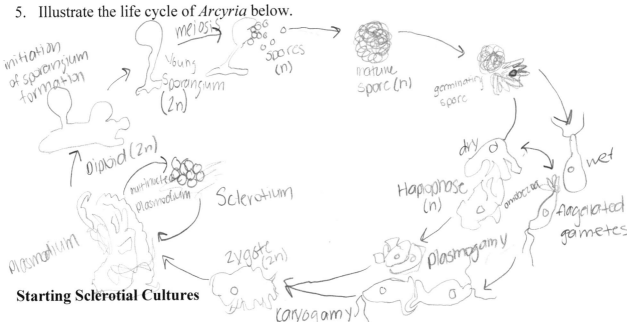

Starting Sclerotial Cultures

Recall that under adverse conditions of temperature and moisture the plasmodium may stop growing and may be converted to a resting stage, the sclerotium. These are hard masses of irregular form consisting of many minute cell-like components. These can however be reactivated to again form growing plasmodia.

- Place a piece of filter paper in a petri dish.

- Wet, but do not soak the filter paper.

- Add of a piece of filter paper containing sclerotia to the surface of the moistened filter paper. Ensure that the filter paper with sclerotia is placed sclerotia side up.

- Keep the filter paper moist over the next couple of days.

- Within a day you should be able to observe active growth of the plasmodium. Place pre-moistened rolled oat flakes in the vicinity of the advancing plasmodial front.

- To make your own sclerotial cultures allow the filter paper to dry then cut into 1 cm squares

Starting Plasmodial Cultures

- Obtain a petri dish with oatmeal agar as well as a swab applicator.

- Use the swab applicator to obtain a portion of the plasmodial culture or spores if these are evident and transfer these to the agar medium.

- Using your sterile (dipped in alcohol) forceps transfer several pre-moistened oat grains to the agar medium.

- Check daily for plasmodial growth.

50

Class: Dictyosteliomycetes (cellular slime molds)

The Dictyosteliomycetes are represented by less than 50 species. Characteristically they have a cell wall and are heterotrophic feeding on bacteria. Cells are free living and ameoboid. Many species, such as those in the genus *Dictyostelium* have a life cycle that is characterized as zygotic meiosis (it only has a free living haploid phase and meiosis occurs after karyogamy in the zygote). Their rather unique asexual life cycle includes the formation of a pseudoplasmodium that is composed of many free living ameoboid cells. These pseudoplasmodia eventually form a sorocarp. This structure is composed of a stalk (sorophore) and an aggregation of spores found at the apex of the stalk. The sorocarp is used in dispersal of the population/colony. Sexual reproduction is not known for most species. The resting structure is a macrocyst.

6. Examine the culture of *Dictyostelium sepium.* Identify the **sorocarp**.

   a. Is this structure part of the **asexual** or **sexual phase** of the life cycle?

   asexual

   b. Is the sorocarp **haploid** or **diploid**?

   haploid

   c. What is the function of the sorocarp?

   It's a mechanism of dispersal of spores in the asexual life cycle

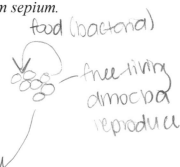
sorocarp

7. Diagram the asexual life cycle in a population of *Dictyostelium sepium.*

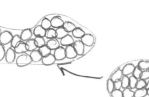

spores
stalk cells

spores
stalk cells

spores

food (bacteria)

free-living amoeba reproduce

Aggregation

51

Class Protosteliomycetes

The Protosteliomycetes are represented by less than 36 species. This group of heterotrophic amoeboid-like organisms that feed on bacteria and fungal cells growing on litter was first described in the 1960s. They are generally microscopic though morphologically diverse and are in part reproductively intermediate or mixed combinations of the myxomycetes or the dictyosteliomycetes. Protostelids are typically only recognized when the amoebae develop into fruiting bodies or sporocarps. The sporocarps consist of little more than a delicate stalk that is typically terminated by a single spore, or depending on the species, possibly 2, 4, or 8 spores.

8. Prepare a culture of protostelids by going to the habitat designated by your instructor and collect a sample of leaf litter. In the lab, use sterile scissors to cut the substrate into small pieces that will be soaked completely in sterile distilled water. Place the leaf litter pieces in a circle on wMY agar. Incubate these plates until the following lab period. Scan the edges of the pieces of leaf litter for protostelid sporocarps the following week. Record your observations.

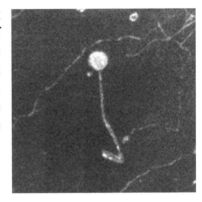

Readings

Baldauf, S. & Doolittle, W.F. 1997. Origin and evolution of the slime molds (Mycetozoa). Proceedings of the National Academy of Sciences of the United States of America 94: 12007-12012.

Gotoh, K. & Kurodo, K. 2005. Motive force of cytoplasmic streaming during plasmodial mitosis of *Physarum polycephalum*. Cell Motility and the Cytoskeleton 2: 173-181.

Kamitsubo, E. & Kikuyama, M. 1994. Bidirectional flow of endoplasm in *Physarum polycephalum* under centrifugal acceleration. Protoplasma 182: 53-58.

Park, D. & Robinson, P.M. 1967. Internal water distribution and cytoplasmic streaming in *Physarum polycephalum*. Annals of Botany 31: 731-738.

Ridgeway, E.B. & Durham, A.C.H. 1976. Oscillations of calcium concentrations in *Physarum polycephalum*. The Journal of Cell Biology 69: 223-226.

Smith, D.A. & Saldana, R. 1992. Model of the Ca2+ oscillator for shuttle streaming in *Physarum polycephalum*. Biophysics Journal 61: 368-380.

Spiegel, F.W., Shadwick, J.D., Lindley, L.L., Brown, M.W., & Nderitu, G. 2007. A beginner's guide to identifying the Protostelids. Fayetteville: University of Arkansas. http://slimemold.uark.edu/pdfs/Handbook1_3rd.pdf Accessed: May 25, 2012.

Domain Eukarya
      Supergroup Opisthokonta
            Phylum Chytridiomycota (Chytrids)

Related Terms: holocarpic (parasitic), eucarpic (parasitic), rhizomycelium (coenocytic - multinucleate), non-nucleate rhizoids, uniflagellate spores, zoosporangia, mitosporangia, isogamy, anisogamy, oogamy, zygote, sporothallus, alternation of generations

Domain Eukarya
      Supergroup Opisthokonta
            Phylum Zygomycota (Bread Molds)

Related terms: mitosporangiophore, mitosporangia, mitospores, gametangia, zygosporangia, conjugation, homothallic, heterothallic, columella

Fungi

Until the 1960s fungi were considered to be members of the Plant Kingdom. Fungi as a group are primary decomposers of organic matter and similar with respect to anatomical and biochemical features. These organisms were placed in their own kingdom with the introduction of the Five-Kingdom system of biological classification. Molecular evidence suggests that the kingdom diverged from other life around 1,500 million years ago. The older, though now less accepted belief was that fungi originally evolved from algae that had lost their chlorophyll. Purportedly the green algae evolved into the much simpler lower fungi (which in this scheme included the Chytridiomycota, Oomycota, and Zygomycota), whereas the red algae evolved into the Ascomycetes. A newer hypothesis proposes that fungi evolved from ancestral flagellates (choanoflagellates) such as algae and protozoa. Today, the hypothesis that fungi evolved from algae, the ancestors to photosynthetic land plants, is not well supported by molecular evidence. Similarly, the hypothesis that fungi evolved independently of both plants and animals is not well supported by molecular evidence. However, that fungi are more closely related to animals than plants is supported by analysis of protein sequence biosynthetic pathways, cytochrome systems, mitochondrial genetic material, biochemical and structural cellular features, glycoproteins, mode of nutrition, and storage of nutritive materials.

Lower Fungi

The lower fungi include the Chytridiomycota, Glomeromycota, and Zygomycota. These three taxa represent the first three lineages of fungi that diverged from the mainstream true fungi. The Chytridiomycota is believed to have diverged first, about 550 million years ago, and as a consequence is more closely allied with animals. Less certainty surrounds the evolutionary history of the Zygomycota. This said, the Zygomycota and the Chytridiomycota were well established by the early Devonian (416-359 Ma) based on the Rhynie chert. Likewise uncertainty surrounds the evolutionary history of the Glomeromycota (arbuscular mycorrhiza) as some data suggests it diverged from the main stream fungi over 460 million years ago when the land flora consisted of only non-vascular bryophyte–like plants.

Phylum Chytridiomycota

This essentially aquatic assemblage occurs in both marine and freshwater environments. The Chytridiomycota are represented by less than 1,000 species of saprobes (organisms that derive nourishment from nonliving organic matter e.g. dead organisms) or parasites (organisms that derive nourishment from living organisms that it lives in or on). Cell walls are composed of chitin (a specific type of carbohydrate) and vegetative hyphae are aseptate (lacking cross walls). The holocarpic/eucarpic chytrids have a free living haploid stage and meiosis occurs

after karyogamy in the diploid zygote. The rhizomycelial groups possess both a free living haploid and diploid generation. These life cycles exhibit zygotic (haplontic) and sporic (haplodiplontic or alternation of generations) meiosis respectively. The life cycle includes both uniflagellate haploid and diploid zoospores and uniflagellate gametes. Chytrids are an environmentally important group as the species *Batrachochytrium dendrobatidis* is believed to be responsible for the worldwide die off of amphibians.

1. Examine the living cultures of *Allomyces*. Make a wet mount of the *Allomyces* gametophyte and sporophyte cultures. Identify **zoosporangia** (asexual), **meiosporangia** (sexual), and any **male** or **female gametangia**. Identify the products of the respective sporangia.

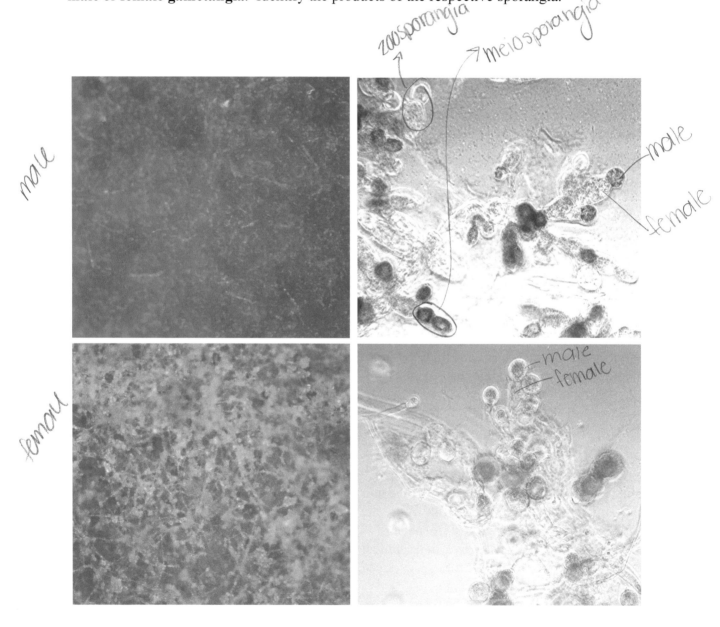

2. Examine slide 99 labeled *Allomyces* gametophyte. Identify the **gametangia**, **gametes**, and **mycelium**. Is the gametophyte of *Allomyces* **holocarpic**, **eucarpic**, or rhizomycelium?

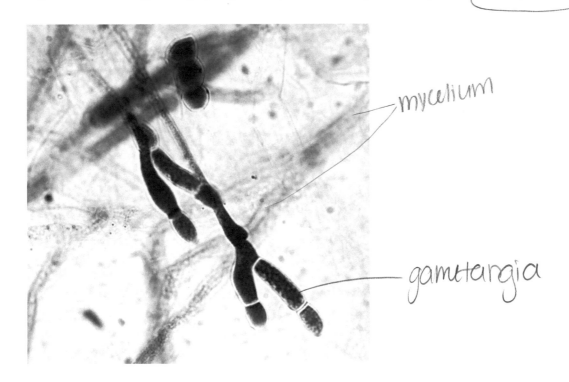

—mycelium

—gametangia

3. Examine slide 100 labeled *Allomyces* sporophyte. Identify the **sporangia, zoosporangia, zoospores, meiospores,** and **mycelium**. Is the sporophyte of *Allomyces* **holocarpic** or **eucarpic**?

mycelium

sporangia

zoosporangia

zoospores

eucarpic

Phylum Zygomycota

The Zygomycota are represented by slightly more than 1,000 species. Cell walls are composed of chitin and the vegetative hyphae are aseptate. This predominantly terrestrial group has only one free living phase in their life cycle, the haploid phase. Thus the life cycle of this predominately terrestrial group is zygotic meiosis or haplontic (meiosis immediately follows karyogamy of the diploid zygote). Sexual reproduction occurs through conjugation and differentiated gametangia occur throughout all members of the phylum. Some species produce gametangia

adorned with sterile appendages. Neither spores nor gametes are flagellate, and the latter are isogamous (the condition where male and female gametes look and behave identically). Members of the phylum are heterotrophic saprobes or parasites. Two of the more well known members of this group include *Rhizopus stolonifer* the black bread mold and *Pilobolus crystallinus* the dung fungus.

4. On demonstration are living cultures of *Mucor* and *Phycomyces blakesleeanus*. Two compatible strains have been inoculated on the plate to induce zygosporangium formation. Zygosporangia form in the region of the darker band. Are these fungi reproducing **sexually** or **asexually**? Are they **homothallic** or **heterothallic**?

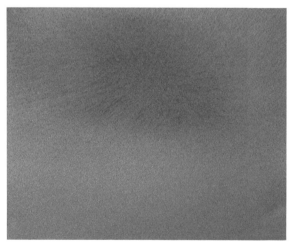
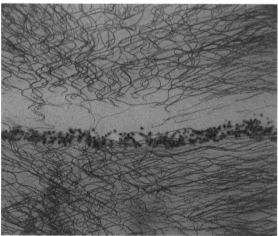

5. Examine slide 6 labeled *Phycomyces.* or prepare a wet mount from a living culture. Draw a **zygosporangium** showing the thick resistant wall supported by the former **conjugating gametangia**, now called **suspensor cells**. Identify **sterile outgrowths (appendages)** and **asexual sporangia**.

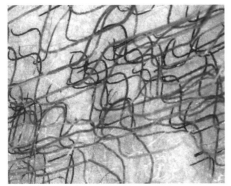

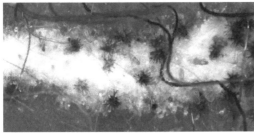
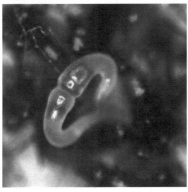

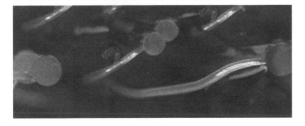

6. Draw and diagram the life cycle of *Rhizopus*. Label where plasmogamy, karyogamy, conjugation and meiosis occur. Identify the **gametangia**, **sporangia**, **zygosporangia**, **suspensor cells**, **mitosporangia**, **mitosporaniophores**, and **hyphae**. Which parts of the life cycle are **haploid** and which are **diploid**?

• two hyphae of diff. mating type grow side by side and produce branches that grow towards each other

• the tips develop into gametangia, structures that produce gametes

• the gametangia, then the gametes will fuze

• the resulting zygote develops into a resistant zygospore

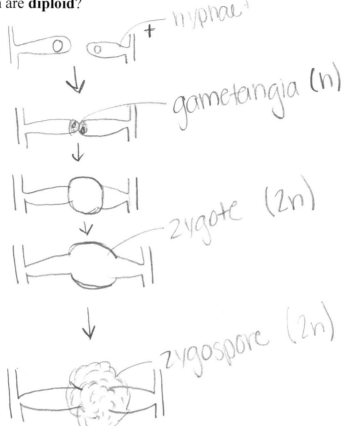

+ hyphae +

gametangia (n)

zygote (2n)

zygospore (2n)

7. Identify three of the major differences among the Myxomycota, Chytridiomycota, and Zygomycota?

1.

2.

3.

8. Prepare a wet mount of the *Rhizopus* culture. Identify the developing **rhizoids, zygosporangia, gametangia, mitosporangia, mitosporangiophore, hyphae, mitospores**, and **columella**.

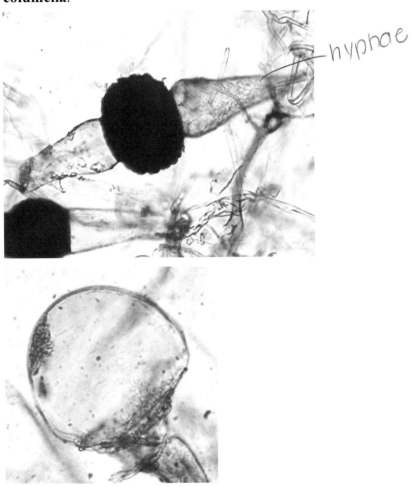

9. Examine slide 9 labeled *Rhizopus* sporangia w.m. and identify the **mitosporangia, mitospores, columella, mitosporangiophore,** and **zygotes**. Which of these structures is **haploid, diploid, part of the sexual phase of the life cycle,** or **part of the asexual phase of the life cycle**?

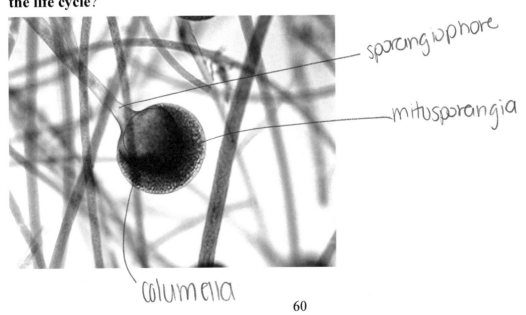

10. Examine slide 13 labeled *Cunninghamella* w.m. and identify the **1-spored mitosporangia** and **branched mitosporangiophore**.

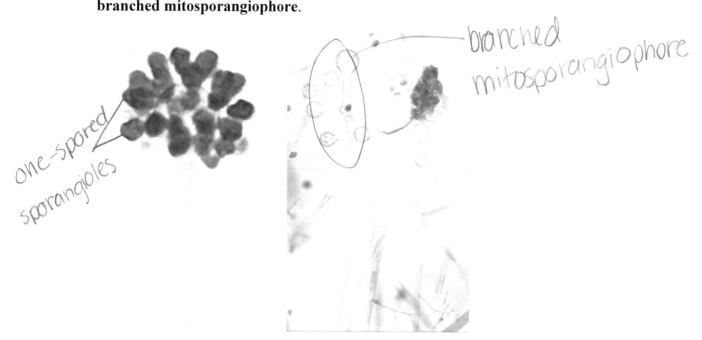

11. Examine slide 14 labeled *Mucor* zygotes w.m. and the live culture. Identify the **zygosporangia, conjugating gametangia, suspensor cells, mitosporangia, hyphae, mitospores, columella,** and **mitosporangiophores**.

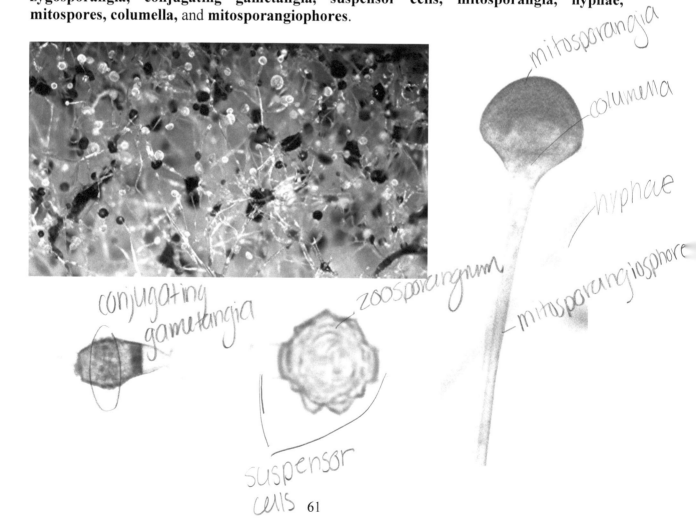

61

12. Examine slide 52 labeled *Thamnidium*.   Identify the **dichotomously branched sporangiophores** and **terminal sporangia**. Do you see **zygosporangia**?

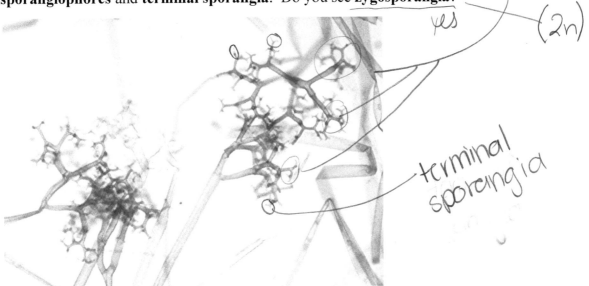

yes   (2n)

terminal
sporangia

13. Examine slide 51 labeled *Entomophthora muscae* on housefly. *Entomophthora muscae* parasitizes houseflies. Masses of sporangiophores bearing sticky apical sporangia which are forcibly shot off at maturity bore through the exoskeleton of the fly. How many masses of sporangiophores occur on your fly? Identify the **sporangia**, **sporangiophores,** and **spores**.

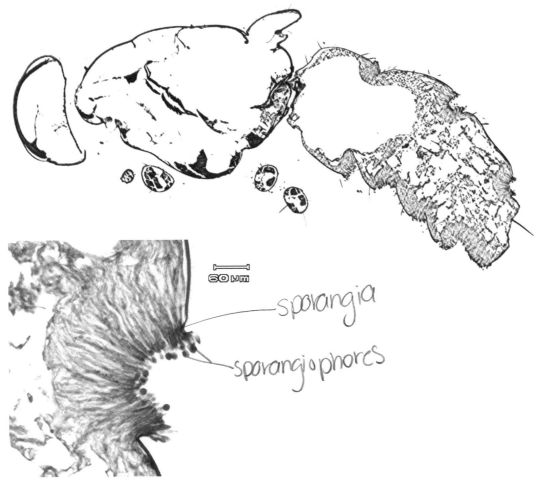

60 μm

sporangia

sporangiophores

14. Prepare a wet mount of the living culture of *Pilobolus*. Identify **sporangiophores**, **sporangia**, **spores,** and **nematodes**.

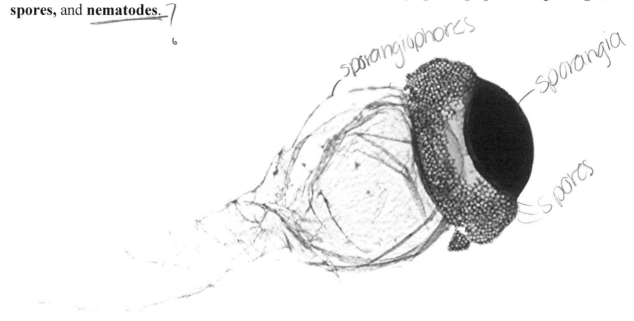

sporangiophores

sporangia

spores

Phylum Glomeromycota

The Glomeromycota are represented by less than 200 species. Their collective importance rests in the symbiotic relationship that they have established with most land plant species. Their coenocytic aseptate hyphae penetrate the cortical cells of plant roots/rhizoids and form arbuscules and vesicles which increase the surface area of the fungus. This facilitates metabolite exchange between the two organisms. The fungus in turn captures nutrients like phosphorus and micronutrients per se which it delivers to the plant. Reproduction is strictly asexual, and sporangia can form inside or outside of the root depending on the species.

15. Examine slide 190 labeled *Corallorrhiza*: rhizome x.s. Identify the **arbuscular mycorrhizal fungi**. In which cell do these fungi occur? Explain.

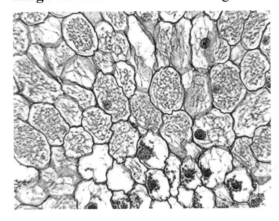

arrow/pink

Readings

Baldauf, S.L. 1999. A search for the origins of animals and fungi: comparing and combining molecular data. American Naturalist 154: S178-S188.

Baldauf, S.L. & Palmer, J.D. 1993. Animals and fungi are each other's closest relatives: congruent evidence from multiple proteins. Proceedings of the National Academy of Sciences of the United States of America 90: 11558-11562.

James, T.Y., Kauff, F., Schoch, C.L., Mathent, P.B., Hofstetter, V., Cox, C.J., Celio, G., Gueidan, C., Fraker, E., Miadlikowska, J., Lumbsch, H.T., Rauhut, A., Reek, V., Arnold, A.E., Amtoft, A., Stajich, J.E., Hosaka, K., Sung, G., Johnson, D., O'Rourke, B., Crockett, M., Binder, M., Curtis, J.M., Slot, J.C., Wang, Z., Wilson, A.W., Schubler, A., Longcore, J.E., O'Donnell, K., Mozley-Standridge, S., Porter, D., Letcher, P.M., Powell, M.J., Taylor, J.W., White, M.M., Griffith, G.W., Davies, D.R., Humber, R.A., Morton, J.B., Sugiyama, J., Rossman, A.Y., Rogers, J.D., Pfister, D.H., Hewitt, D., Hansen, K., Hambleton, S., Shoemaker, R.A., Kohlmeyer, J., Volkmann-Kohlmeyer, B., Spotts, R.A., Serdani, M., Crous, P.W., Hughes, K.W., Matsuura, K., Langer, E., Langer, G., Untereiner, W.A., Lucking, R., Budel, B., Geiser, D.M., Aptroot, A., Diederich, P., Schmitt, I., Schultz, M., Yahr, R., Hibbett, D.S., Lutzoni, F., McLaughlin, D.J., Spatafora, J.W., & Vilgalys, R. Reconstructing the early evolution of fungi using a six-gene phylogeny. 2006. Nature 443: 818-822.

Martin, G.W. 1955. Are fungi plants? Mycologia 47: 779-791.

Redecker, D. 2002. New views on fungal evolution based on DNA markers and the fossil record. Research in Microbiology 153: 125-130.

Tunlid, A. & Talbot, N.J. 2002. Genomics of parasitic and symbiotic fungi. Current Opinion in Microbiology 5: 513-519.

Wang, D.Y.C., Kumar, S., & Hedges, S.B. 1999. Divergence time estimates for the early history of animal phyla and the origin of plants, animals, and fungi. Proceedings of the Royal Society of London. Series B 266: 163-171.

Domain Eukarya
    Supergroup Opisthokonta
        Phylum Ascomycota (Sac Fungi)
            Subphylum Pezizomycotina
                Class Discomycetes
                Class Plectomycetes
                Class Pyrenomycetes
                Class Loculoascomycetes
            Subphylum Hemiascomycotina (*sensu* Brefeldt 1891)

Related terms: ascoma, ascogonium, antheridium, trichogyne, dikaryon, ascogenous hyphae, hymenium, crozier, karyogamy, endogenous spore development, unitunicate, operculum, operculate asci, ascospores, apical ring apparatus, bitunicate, ascus, apothecial ascoma, perithecial ascoma, stroma, cleistothecial ascoma, pseudothecial ascoma, paraphyses, periphyses, pseudoparaphyses, teleomorph, anamorph, haustoria

Higher Fungi I

The higher fungi, which include the Ascomycota, Basidiomycota and Deuteromycota, represent that group of fungi which diverged most recently from the main stream fungi lineage. The Ascomycota and Basidiomycota diverged from one another about 400 Ma and all modern groups were present by the end of the Carboniferous Period, specifically the Pennsylvanian Epoch (318-299 Ma).

Phylum Ascomycota

The Ascomycota, also known as the "sac fungi", are represented by more than 64,000 species in two Subphyla. Those in Subphylum Pezizomycotina (Euascomycotina) are hyphal and produce a fruiting body (ascoma), whereas those in Subphylum Hemiascomycotina do not produce ascoma. Some of the more notable members of the former group include morels (*Morchella*) and truffles (*Tuber*) both of which are gastronomic delicacies. The haploid hyphae of the Ascomycota are composed of chitin and are septate. Septa in the Pezizomycotina are associated with Woronin bodies (membrane bound structures that are composed of proteins and lipids that seal the septal pore if the hyphal strand is damaged). The most visible part of the life cycle (as they are visible to the unaided eye) in the

Ascomycota are the sexual reproductive structures known as ascoma (ascocarp, ascomata pl.). Four types of ascoma are produced by the haploid hyphae, apothecia, cleistothecia, perithecia, and pseudothecia. Gametangia within form both antheridia (♂) and ascogonia (♀) which come together via an outgrowth (trichogyne) from the latter and result in the formation of the dikaryon. Ascogenous dikaryotic hyphae grow from the ascogonium and eventually form a sac-like structure (ascus) which may be one (unitunicate) or two (bitunicate) layers thick within which haploid ascospores are produced (i.e. endogenous spores). Most species produce 8 ascospores per ascus while some species have only one and others up to 1,024. The sexual phase in the Ascomycota is referred to as the teleomorphic life cycle while the asexual phase is called the anamorphic life cycle. Specialized sterile hyphae may be associated with the asci. These include: paraphyses, periphyses, and pseudoparaphyses. The Ascomycota exhibits the zygotic meiosis life cycle (meiosis immediately follows karyogamy of the diploid zygote). In almost all cases, and gametes are oogamous-like (female gametes are sessile while the male gametes are motile). Odd as it may sound, asexual reproduction in the ascomycetes will be discussed with the Deuteromycota.

Phylum Ascomycota
    Subphylum Pezizomycotina (Euascomycotina)
        Class Discomycetes: apothecia with unitunicate asci
        Class Plectomycetes: cleistothecia with unitunicate asci
        Class Pyrenomycetes: perithecia with unitunicate asci
        Class Loculoascomycetes: pseudothecia with bitunicate asci
    Subphylum Hemiascomycotina

Terms:

apothecia - cup-like with an exposed fertile layer

cleistothecia - a closed spherical structure with a fertile layer

perithecia - a hollow flask-shaped chamber with an opening at the top and a fertile layer

pseudothecia - a locule within a mass of hyphae containing an irregular arrangement of asci.

1. Examine slide 16 of *Eurotium*, which is incorrectly labeled *Aspergillus* cleistothecia. *Eurotium* is the teleomorph phase of the life cycle and *Aspergillus* is the anamorph. Identify **asci** and **ascospores** within the **cleistothecia**. To which Class does *Eurotium* belong.

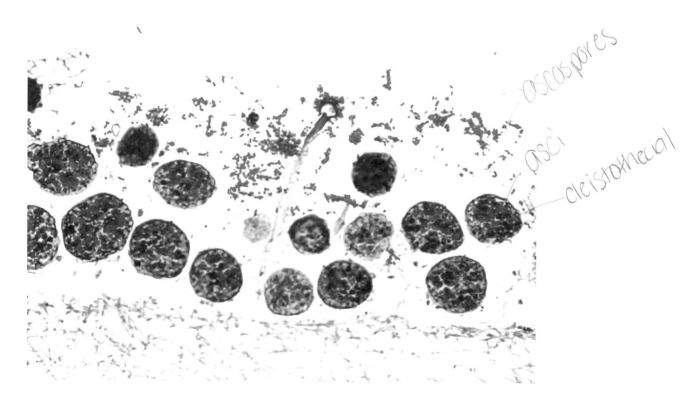

2. Examine slide 19 of *Eupenicillium* which is incorrectly labeled *Penicillium* with cleistothecia. Identify **asci**, **ascospores,** and **cleistothecia**.

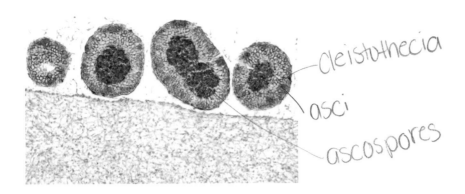

cleistothecia

asci

ascospores

3. Examine slide 22 labeled *Ceratocystis* sp. Note the **hyphal** impregnations in the xylem.

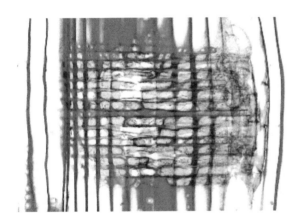

4. Examine slide 23 labeled *Microsphaera* asci – in some cases these are whole mounts in other cases they are sectioned - observe both types. Note the dichotomously branched appendages in the whole mounts as well as haustoria in the sectioned mounts - you will probably have to examine several slides to find a good example of each! How do rhizoids differ from haustoria?

powdery mildews

haustoria

rhizoid

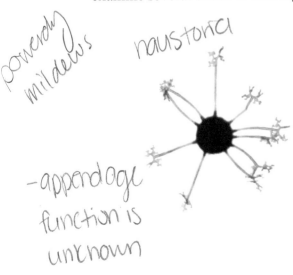

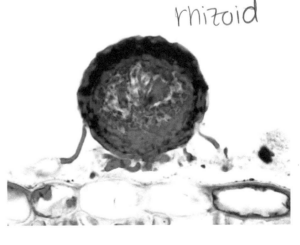

-appendage function is unknown

-function of appendages is to anchor

5. Examine slide 25 labeled *Uncinula salicis* cleistothecia. Note the **curled appendages**, **asci**, **ascospores**, and **haustoria** - you may have to examine several slides to find good examples!

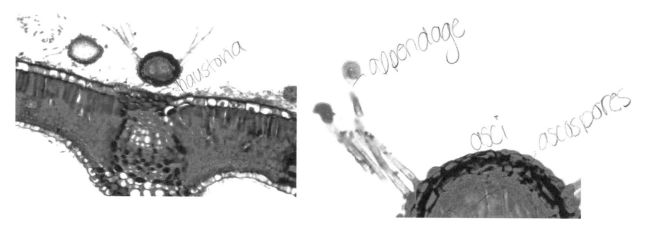

6. Examine slide 26 labeled *Phyllactinia corylea.* Note the **needle-like appendages**, **asci**, **ascospores**, and **haustoria** - you may have to examine several slides to find good examples!

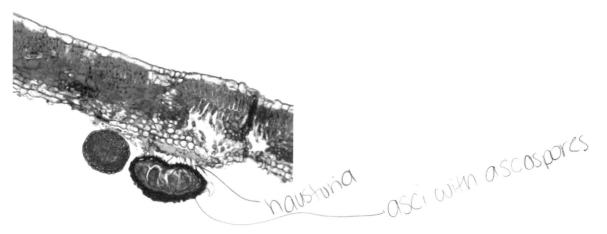

7. Examine slide 28 labeled *Monilinia fructicola* apothecium. Identify the **stalked ascoma, asci, ascospores,** and **hymenium**.

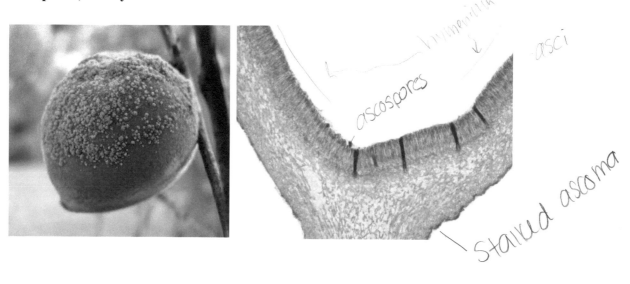

8. Examine slide 29 labeled *Peziza* apothecia. Identify the **ascoma, asci, ascospores,** and **hymenium**.

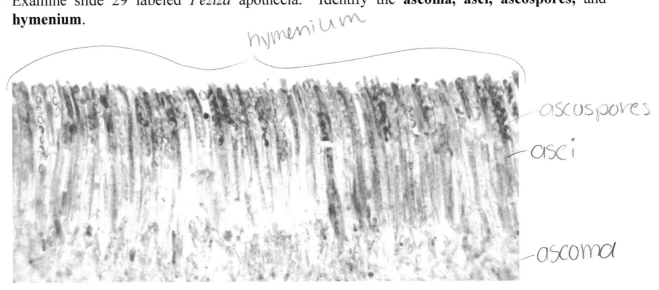

hymenium

ascospores

asci

ascoma

9. Examine slide 31 labeled *Erysiphe graminis* cleistothecia. Identify the **ascoma, asci, ascospores,** and **haustoria**. Are the asci **unitunicate** or **bitunicate**?

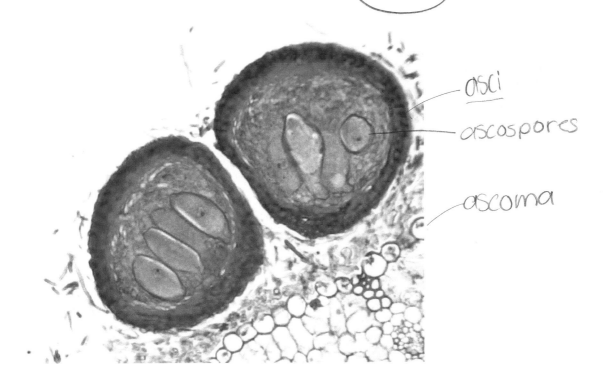

asci

ascospores

ascoma

10. Examine slide 21 labeled *Venturia inequalis* ascocarp. Identify the type of **ascoma, pseudoparaphyses, asci,** and **ascospores.** Are the asci **unitunicate** or **bitunicate**?

elongated cells → not an ascus!

pseudothecial ascoma

ascospores

asci

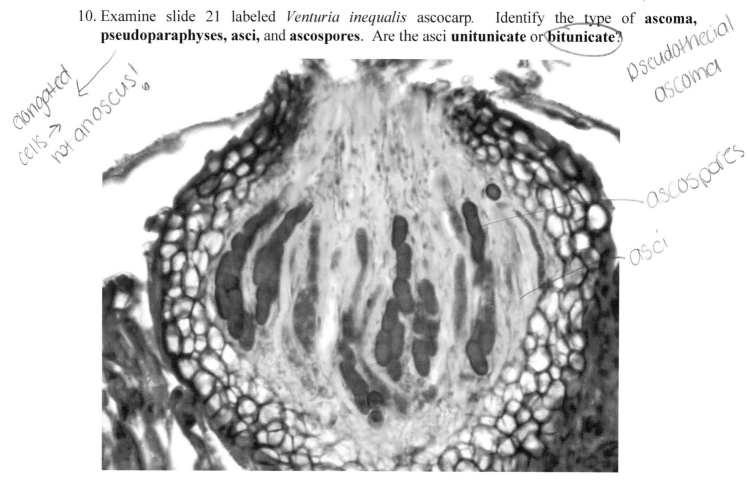

11. Examine slide 53 labeled *Ergot* mature stroma which is *Claviceps purpurea.* Note the **stalked stroma** with spherical head within which develop many **perithecia.** Identify the **neck, ostiole, periphyses, paraphyses, asci,** and **ascospores.** Is this the **teleomorph** or **anamorph**?

 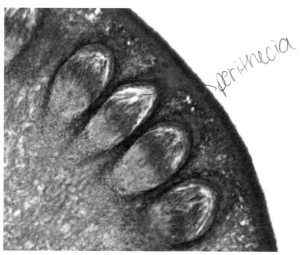

perithecia

71

12. Examine the preserved specimen of *Morchella* on demonstration. Note the ridges and pits. What term best applies to the fruiting body.

13. Examine slide 50 labeled *Morchella*. Identify the **asci** and **ascospores**.

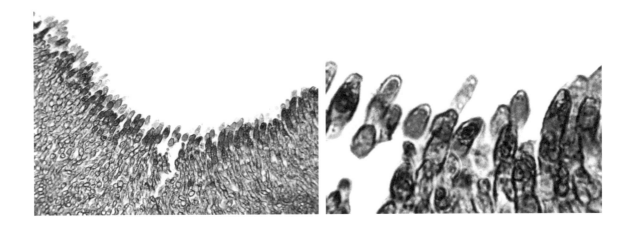

14. Examine slide 57 labeled *Xylaria* perithecia. Identify the **stroma**, **perithecia, ostiole, paraphyses, periphyses**, **asci,** and **ascospores.**

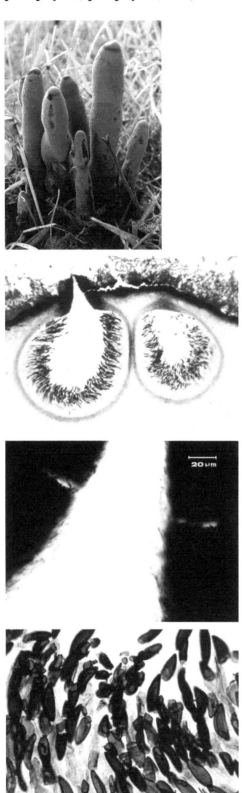

15. Examine slide 59 labeled *Schizosaccharomyces*. The yeast on this slide has undergone sexual reproduction. Find an **ascus (meiosporangium)** containing the **ascospores**. How many ascospores are produced? Are yeasts typically **haploid** or **diploid**? Remember that while *Schizosaccharomyces* produces asci, some other yeast produce basidia. This means that their simple morphology is misleading and they are a heterogeneous polyphyletic group.

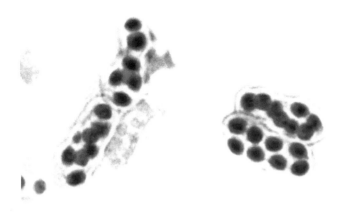

16. Examine slide 8 labeled *Saccharomyces* vegetative. Identify budding cells.

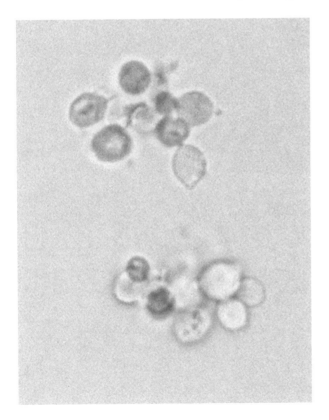

17. Diagram and label the life cycle of a yeast (Hemiascomycotina). Where do **plasmogamy, karyogamy,** and **meiosis** occur? Identify **haploid, diploid,** and **dikaryotic regions** in the yeast life cycle.

18. Diagram and label the life cycle of a typical ascomycete. Where do **plasmogamy, karyogamy,** and **meiosis** occur? Identify and label **ascogonia, antheridia, gametangia, sporangia,** and **monokaryotic cells**. Identify **haploid, diploid,** and **dikaryotic regions** in the ascomycete life cycle.

19. Prepare a spore print for next week using the club fungus provided by your instructor.

Readings

Demoulin, V. 1974. The origin of Ascomycetes and Basidiomycetes. The case for a red algal ancestry. Botanical Review 40, 315-345.

Dhavale, H. & Jedd, G. 2007. The fungal Woronin body. *In*: Biology of the fungal cell VIII. The Mycota 8:87-96.

Kohlmeyer, J. 1975. New clues to the possible origin of the Ascomycetes. BioScience 25: 86-93.

Schultz, M., Arendholtz, W.-R., & Büdel, B. 2008. Origin and evolution of the lichenized ascomycete Order Lichinales: Monophyly and systematic relationships inferred from ascus, fruiting body and SSU rDNA evolution. Plant Biology 3: 116-123.

V.H.B. 1903. The origin of the Ascomycetes. New Phytologist 2:112-114.

Liu, Y-.J.J. & Hall, B.D. 2004. Body plan evolution of ascomycetes, as inferred from an RNA polymerase II phylogeny. Proceedings of the National Academy of Sciences of the United States of America 101: 4507-4512.

Domain Eukarya
    Supergroup Opisthokonta
        Phylum Basidiomycota (Club Fungi)
            Class Holobasidiomycetes
                Order Agaricales (Hymenomycetes)
                Order Polyporales (Hymenomycetes)
                Order Lycoperdales (Gasteromycetes)
                Order Sclerodermatales (Gasteromycetes)
                Order Nidulariales (Gasteromycetes)
                Order Phallales (Gasteromycetes)
            Class Phragmobasidiomycetes
                Order Tremellales (cruciately septate)
                Order Auriculariales (transversely septate)
            Class Teliomycetes
                Order Ustilaginales (Smuts)
                Order Uredinales (Rusts)

Related terms: monokaryon, clamp connection, dikaryon, basidioma, sterigmata, basidiospore, hymenium, agarics, gills, lamellae, stipe, stalk, cap, pileus, universal veil, volva, partial veil, ring, annulus, peridium, tubes, haustoria, heterocious, alternate host, primary host, spermagonia, spermatia, aecia, transfer spores, aeciospores, uredinial sorus, uredinium, urediniospores (technically the anamorph stage), summer spores, infection stage, telial sorus, teliospores, winter spores, bicellular, abaxial, adaxial

Higher Fungi II

Phylum Basidiomycota

The Basidiomycota, also known as the "club fungi" are a terrestrial group that is represented by nearly 30,000 species. The most conspicuous part of a basidiomycete is the fruiting body (basidioma, basidiocarp). The Class and Order assemblages within this group are based on characteristics associated with the basidium and basidioma, respectively (see below). Although the most conspicuous feature of the basidiomycetes is the basidoma the most dominant feature is the mycelium. Cell walls are composed of chitin and the vegetative hyphae have incomplete septa that are associated with parenthosomes (sometimes the entire complex is referred to as a dolipore plug). The mycelium that forms the basidioma is predominately dikaryotic. Maintenance of the dikaryon is ensured by the clamp connections.

There are no differentiated gametangia and meiospores (basidiospores) form exogenously. Basidiospores are ultimately derived from a basidium (a structure in which karyogamy and meiosis takes place). The Basidiomycota exhibit the zygotic meiosis life cycle (meiosis immediately follows karyogamy of the diploid zygote) and gametes are isogamous (the condition where male and female gametes look and behave identically).

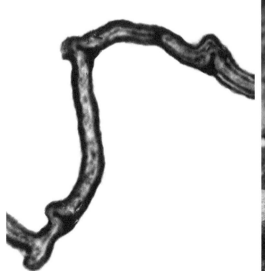

78

Phylum Basidiomycota
    Separation of classes based on basidium types

        Class Holobasidiomycetes (holobasidium)
            Order Agaricales   **"Hymenomycete group"**
                Family Agaricaceae
                Family Boletaceae
            Order Polyporales (Aphyllophorales)
            Order Lycoperdales   **"Gasteromycete group"**
            Order Niduriales
            Order Phallales
            Order Sclerodermatales
        Class Phragmobasidiomycetes (phragmobasidium)
            Order Auriculariales (transversely septate)
            Order Tremellales (cruciately septate)
        Class Teliomycetes (teliobasidium)
            Order Urediniales
            Order Ustilaginales

Terms:

holobasidium - a one celled basidium

phragmobasidium - a septate basidium

teliobasidium - a thick walled resting spore with a promycelium (functionally a series of basidia) that bears the basidiospores

hymenomycete group – basidiospores are directly exposed to the external environment

gasteromycete group – basidiospores are enclosed in 1-2 protective layers called a peridium.

1. Examine the spore print you prepared in lab last week. Note the spore pattern as well as spore color. How does your spore print compare with the one below?

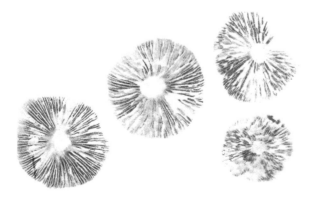

2. Investigate the edible mushroom *Agaricus brunnescens*. Cut a specimen in half longitudinally and examine the **cap**, **stalk**, **partial veil**, **gills**, and **hymenium**.

How is the partial veil formed?

Where are the spores?

What color are the spores?

How are the spores dispersed?

3. Examine slide 44 labeled *Coprinus* pileus c.s. Identify the **stipe**, **cap**, **hymenium**, **hyphae**, **basidia**, **sterigmata**, and **basidiospores**. Is this a (**hymenomycete**) or a **gasteromycete**? Explain.

4. Examine slide 42 labeled *Boletus* pores c.s. Identify the **basidia**, **pores/tubes**, **hymenium**, and **basidiospores**.

hymenium

pores

hymenium

basidiospores

5. Examine slide 48 labeled *Polyporus* pores c.s. Identify the **pores/tubes**, **hymenium**, **basidia**, and **basidiospores**. How do the pores/tubes of Order Agaricales Family Boletaceae compare with those of Order Polyporales? Explain.

these ones
attach from each other

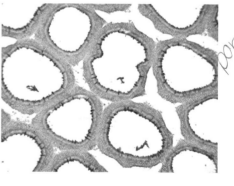

pores

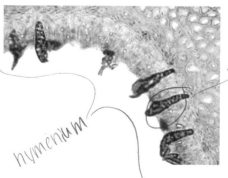

basidia
–basidiospores

hymenium

hymenomycetes

6. Examine slide 43 labeled *Clavaria*. This fungus belongs in Class Holobasidiomycetes, Order Polyporales and is therefore included in the hymenomycete group. *Clavaria* is atypical in that the basidia are produced terminally on the branches of the basidoma as opposed to in tubes or gills. Identify the **hymenium**, **hyphae**, **basidia**, and **basidiospores**.

7. Examine the demonstration material for Order Lycoperdales (top image) and Order Sclerodermatales (bottom image). How do members of these two Orders differ? Explain.

Lycoperdales Order
- true puffballs
- ostiole

Sclerodermatales Order
- false puffballs

8. Examine slide 45 labeled *Cyathus*. This is a longitudinal section of the basidioma. Identify the **peridium, peridicles, peridioles**, and **basidiospores**. Is this a **hymenomycete** or **gasteromycete**? Explain.

peridicles
→atochny
in peridioles
enclosed
↓?
peridioles
peridioles

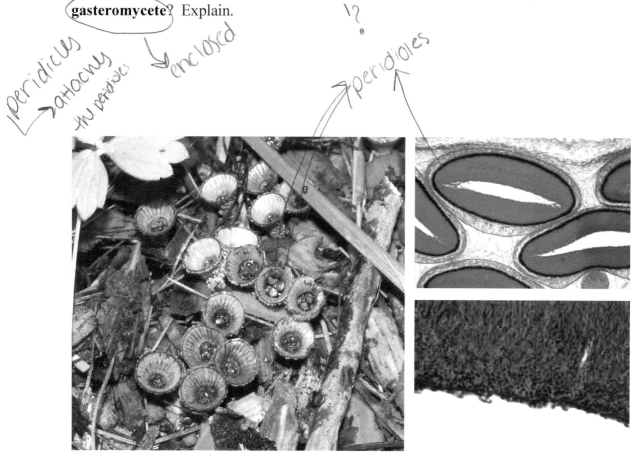

9. Examine slide 46 labeled *Geaster* section. Identify the two layers of the **peridium**. Where are the **basidiospores** located? Do you see an ostiole?

outer peridium

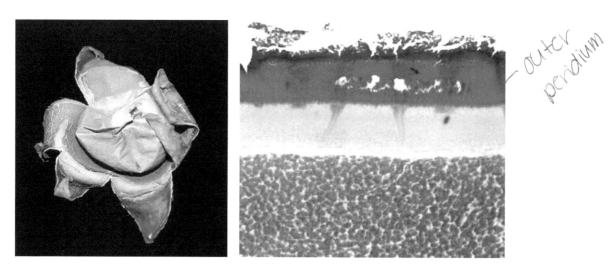

83

10. Examine slide 47 labeled *Mutinus caninus* fruiting body. *Mutinus caninus* is the dog stinkhorn. Identify the **peridium**, **hollow stalk**, and **basidiospores**.

hollow stalk

peridium

spores somewhere

11. Examine slide 49 labeled *Tremella* section. *Tremella* belongs to the Order Tremellales, the jelly fungi, usually found on decaying wood. The basidium becomes vertically divided into four cells after nuclear fusion and meiosis have occurred. From each of the four compartments arises a long hyphal process with a terminal **sterigma** from which a basidiospore is produced.

↳ circle growths

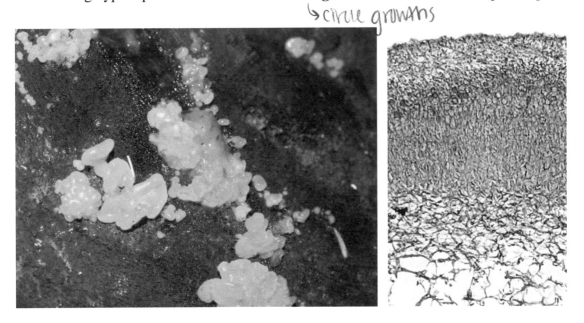

12. Members of the Urediniales (rusts) have a complicated life cycle that involves two hosts (heterocious) and up to five developmental stages. The host of wheat rust includes barberry and as the name suggests wheat. Rusts like other members of Phylum Basidiomycota exhibit zygotic meiosis, but differ in having several asexual phases.

13. Examine slide 36 labeled *Puccinia graminis* **aecia** and **pycnia**. Basidiospores which land on the adaxial surface of barberry leaves germinate to produce pycnia (spermagonia) from which receptive hyphae grow and spermatia ooze. When compatible receptive hyphae and spermatia come in contact with each other the dikaryon is established. The dikaryotic hyphae then grow intercellularly through the mesophyll to the abaxial side of the leaf. Identify the reproductive structures as well as their contents. Are the reproductive structures being produced on the **adaxial** or **abaxial surface of the leaf**? Identify the host. Are these cells **haploid** or **diploid**? Are these cells **uninuclelate** or **binucleate**?

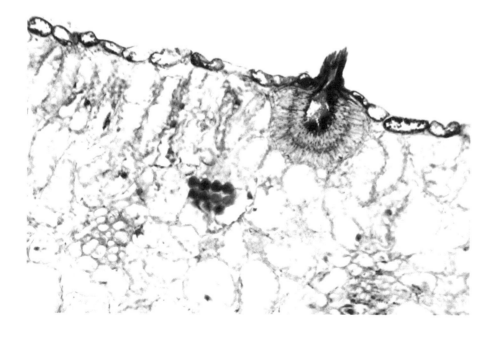

14. Examine slide 35 labeled *Puccinia graminis* **aecia**. The dikaryotic hyphae which have grown to the abaxial surface of the leaf differentiate to form aecia and chains of dikaryotic aeciaospores. The aeciaospores, also known as transfer spores are wind dispersed and infect wheat (the primary host). Identify the structure within which the spores are being produced. What type of spores are these? Are the reproductive structures being produced on the **adaxial** or **abaxial surface of the leaf**? Identify the host organism.

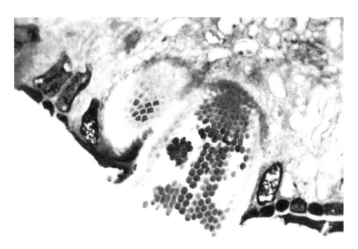

15. Examine slide 34 labeled *Puccinia graminis* **uredinia**. The dikaryotic uredinial spores are reddish and produced in clusters. When released the uredinial spores infect wheat plants. This sequence of release and infection is repeated throughout the growing season. Are these cells **haploid** or **diploid**? Are these cells **uninuclelate** or **binucleate**? Which stage in the life cycle of the wheat rust are you viewing? Identify the host organism.

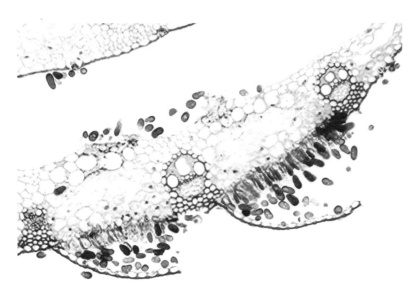

16. Examine slide 33 labeled *Puccinia graminis* telia. Note the two-celled **teliospores**. The teliospore represents the resting spore. Are these cells **haploid** or **diploid**? Which stage in the life cycle occurs below? Identify the host organism.

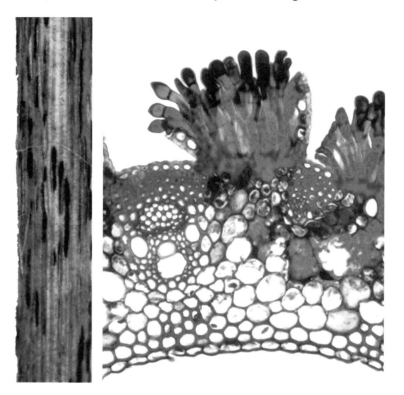

17. Examine slide 40 labeled *Cronartium ribicola* spermagonia. *Cronartium ribicola* is the causal agent of white pine blister rust. Both spermatia and aeciospores are produced on *Pinus strobus* and the urediniospores and teliospores are produced on *Ribes* (currants/gooseberries). The teliospores germinate soon after they are produced. The basidiospores infect pine and the fungus overwinters as mycelium in the pine. Identify the **spermatia** and **spermagonia**.

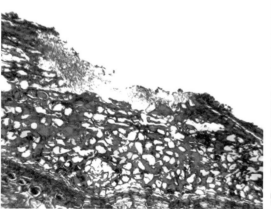 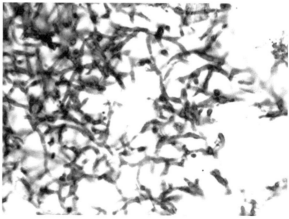

18. Examine slide 39 labeled *Cronartium ribicola* aecia. Identify the **aecium** and **aeciospores**. Compare/contrast the aecium of *Cronartium* with *Puccinia*.

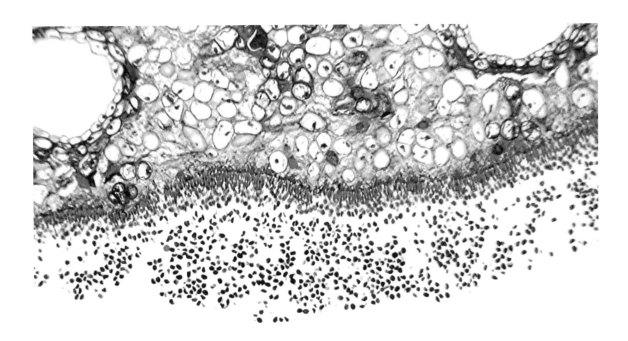

19. Examine slide 37 labeled *Cronartium ribicola* uredinia. Are these cells **haploid** or **diploid**? Are these cells **uninucleate** or **binucleate**? Which stage in the life cycle are you viewing? Identify the host organism.

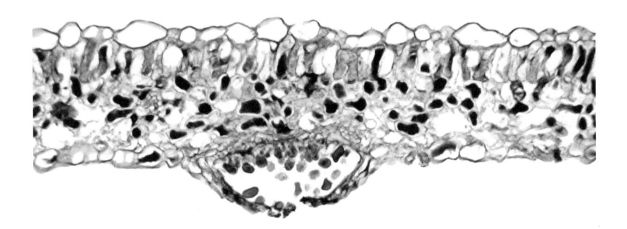

20. Examine slide 38 labeled *Cronartium ribicola* telia. Identify the reproductive structures. Identify the host organism. Note the nuclei!

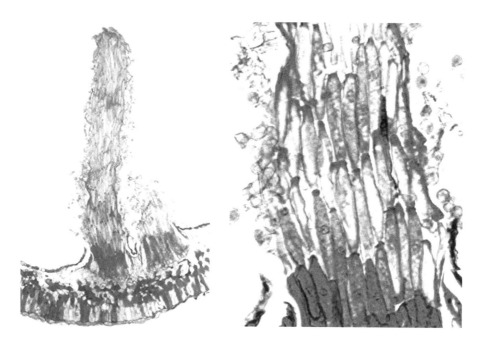

21. Examine slide 41 labeled *Ustilago zeae* or *maydis* (in corn). The teliospore of a smut is karyologically equivalent to that of a rust. After germination, the hypha becomes septate and buds off a basidiospore from each cell. The basidiospore of smut fungi often infects the host ovary during flowering and produces a haploid mycelium which soon fuses with another compatible monokaryon to establish the dikaryon. The binucleate mycelium finally produces a mass of teliospores (smut spores) where the new seed would have been. Identify the **smut spores**.

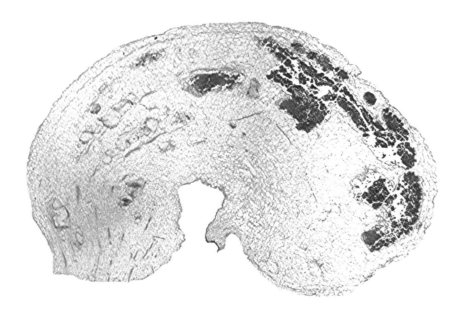

22. Diagram and label the life cycle of a typical wheat rust. Where do **plasmogamy, karyogamy, and meiosis** occur? Identify and label **haploid, diploid,** or **dikaryotic regions**. Identify and label **aecia, aeciospores, spermatia, spermagonia, uredinia, uredinia spores, telium,** and **teliospores.**

23. Diagram and label the life cycle of the mushroom *Agaricus*. Label where **plasmogamy, karyogamy,** and **meiosis** occur. Identify and label where it is **haploid** or **diploid**. Identify **gametangia, sporangia, monokaryotic cells**, and **dikaryotic cells**.

Readings

Alexopoulos, C. J. 1996. Introductory Mycology. John Wiley & Sons, New York.

Arora, D. 1986. Mushrooms Demystified, A Comprehensive Guide to the Fungi. Ten Speed Press, Berkley, CA.

Bessette, A.E., Bessette, A.R., & Fischer, D.W. 1997. Mushrooms of Northeastern North America. Syracuse University Press, Syracuse, NY.

Findlay, W.P.K. 1967. Wayside and Woodland Fungi. Frederick Warne & Co. Ltd., London, England.

Fischer, G.W. & Holton, C.S. 1957. Biology and control of the smut fungi. The Ronald Press Company, New York.

Graham, V.O. 1970. Mushrooms of the Great Lakes region; the fleshy, leathery, and woody fungi of Illinois, Indiana, Ohio, and the southern half of Wisconsin and of Michigan, Dover Publications, New York.

Hibbett, D.S., Binder, M., Bischoff, J.F., Blackwell, M., Cannon, P.F., Eriksson, O.E., Huhndorf, S., James, T., Kirk, P.M., Lücking, R., Thorsten Lumbsch, H., Lutzoni, F., Matheny, P.B., McLaughlin, D.J., Powell, M.J., Redhead, S., Schoch, C.L., Spatafora, J.W., Stalpers, J.A., Vilgalys, R., Aime, M.C., Aptroot, A., Bauer, R., Begerow, D., Benny, G.L., Castlebury, L.A., Crous, P.W., Dai, Y.C., Gams, W., Geiser, D.M., Griffith, G.W., Gueidan, C., Hawksworth, D.L., Hestmark, G., Hosaka, K., Humber, R.A., Hyde, K.D., Ironside, J.E., Kõljalg, U., Kurtzman, C.P., Larsson, K.H., Lichtwardt, R., Longcore, J., Miadlikowska, J., Miller, A., Moncalvo, J.M., Mozley-Standridge, S., Oberwinkler, F., Parmasto, E., Reeb, V., Rogers, J.D., Roux, C., Ryvarden, L., Sampaio, J.P., Schüssler, A., Sugiyama, J., Thorn, R.G., Tibell, L., Untereiner, W.A., Walker, C., Wang, Z., Weir, A., Weiss, M., White, M.M., Winka, K., Yao, Y.J., & Zhang, N. 2007. A higher level phylogenetic classification of the Fungi. Mycological Research 111: 509–547.

Marshall, N.L. 1923. A Popular Guide to the Identification and Study of Our Commoner Fungi, with Special Emphasis on the Edible Varieties. Garden City, N. Y., Doubleday, Page and company.

Moore, R.T. 1980 Taxonomic proposals for the classification of marine yeasts and other yeast-like fungi including the smuts. Botanica Marina 23: 371.

Domain Eukarya
    Supergroup Opisthokonta
        Phylum Deuteromycota (Fungi Imperfecti)
           Class Coelomycete
           Class Hyphomycete

Related terms: conidial fungi, conidia, conidiophores, coelomycetes (form conidioma), pycnidium, acervulus, hyphomycetes (no conidioma), amerospore, didymospore, helicospore, staurospore, dictyospore, phragmospore, scolecospore, retrogressive blastic development, basauxic blastic development, thallic development

Higher Fungi III

Phylum Deuteromycota

The Deuteromycota also known as the fungi imperfecti produce conidial spores through asexual reproduction (anamorph phase). As we have seen in the previous two labs both the ascomycetes and basidiomycetes are classified based on characteristics of the fruiting body (ascoma and basidioma, respectively) and spores (ascospores and basidiospores, respectively) produced within these during sexual reproduction (teleomorph). Historically a dual system of nomenclature following Article 59 of the International Code of Botanical Nomenclature (adopted in the Brussels Botanical Congress of 1912) was used to

distinguish between related anamorphs and teleomorphs. For this reason the Deuteromycota were placed in their own phylum (division) within the Kingdom Fungi. Their classification was further refined based on whether the conidiospores were naked (Class Hyphomycetes) or produced within fruiting bodies (Class Coelomycetes). This dual form of nomenclature has been and continues to be controversial.

Phylum Deuteromycota

Class Coelomycete: conidia develop inside some type of protective structure (**conidiomata**)

Class Hyphomycete: conidia do not develop in a protective structure

1. Examine slide 20 labeled *Venturia inequalis* conidia. In reality, the anamorph of the apple scab fungus is *Spilocaea pomi*. Identify the **conidia** and **conidiophores**.

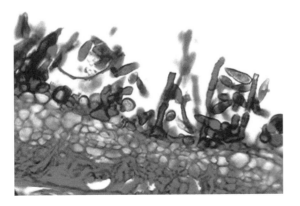

2. Examine slide 32 labeled *Helminthosporium maydis* on *Zea*. This fungus is responsible for the southern corn blight which devastated the U.S. corn crop in 1970. Identify the **conidia** and **conidiophores.** How do these conidia compare with those of *Venturia inequalis*.

3. Examine slide 30, labeled *Erysiphe graminis* conidiophores. *Erysiphe* is a powdery mildew which produces whitish chains of conidia over the leaves. Each chain consists of a graded series of gradually maturing conidia, the oldest at the tip, the youngest, barely differentiated from the hyphal cell just below it. Identify the **conidia** and **conidiophores**. What term best describes this type of conidial development? To which class does *Erysiphe graminis* belong?

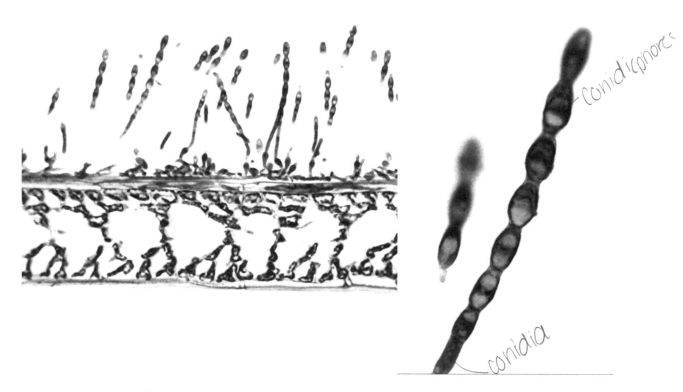

95

4. Examine slide 24 labeled *Microsphaera alni* conidia. Identify **conidia** and **conidiophores**.

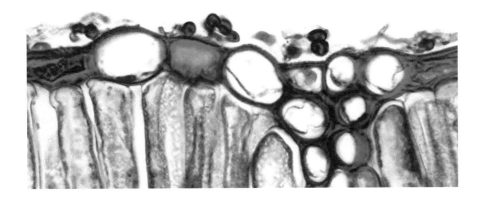

5. Examine slide 27 labeled *Monilinia fructicola* conidia. Identify **conidia** and **conidiophores**. What term best describes this type of conidial development?

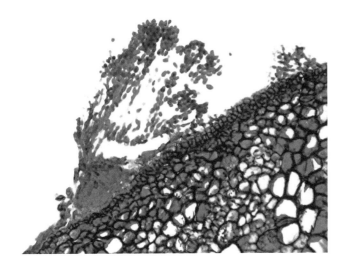

6. Examine the live culture of *Aspergillus niger*. What is the relationship between the color of the colony and the specific epithet? Prepare a wet mount of the colony. Draw. Identify **hyphae**, **conidia**, **conidiophores**, and **conidiogenous cells**.

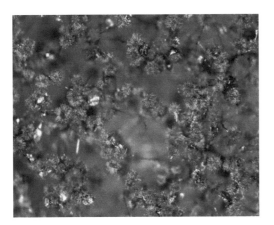

7. Examine slide 15 labeled *Aspergillus*. Identify **hyphae**, **conidia**, **conidiophores**, and **conidiogenous cells**.

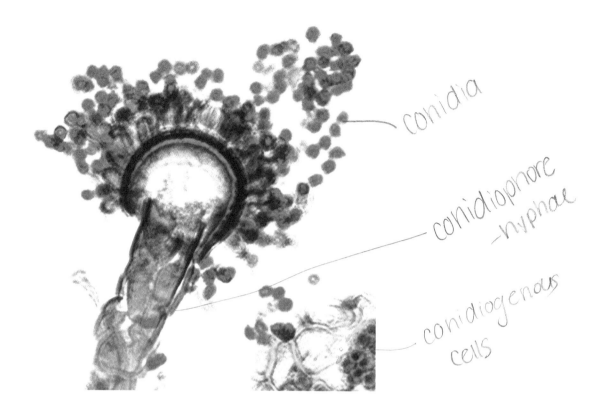

conidia

conidiophore

hyphae

conidiogenous cells

8. Examine the live culture of *Penicillium roqueforti*. Have you ever eaten this fungus in the past? Prepare a wet mount of the colony. Draw. Identify **hyphae**, **conidia**, **phiallides conidiophores**, and **conidiogenous cells**.

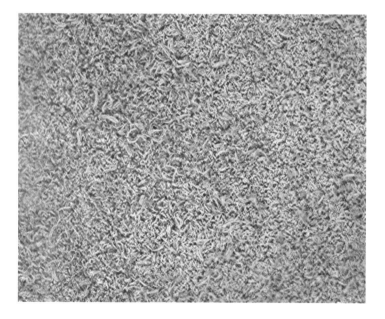

9. Examine slide 18 labeled *Penicillium* with conidiophores. Identify **conidia**, **hyphae**, **conidiophores**, **phiallides**, and **conidiogenous cells**.

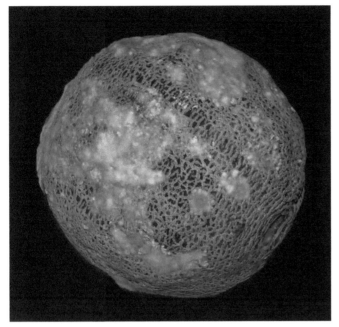

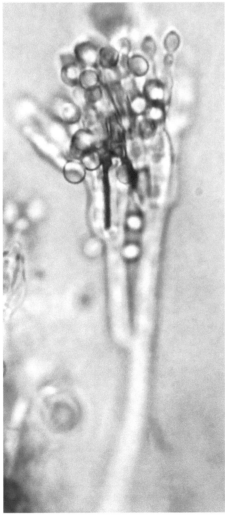

10. Examine slide 17 labeled mold types. **Identify** the respective reproductive structures for *Aspergillus, Penicillium*, and *Rhizopus* (i.e. **conidiophores, sporangiophores, etc.**).

    a. What is the difference between the arrangement of asexual spores in *Rhizopus* as compared to *Aspergillus* and *Penicillium*?

    b. To which phylum does *Rhizopus* belong?

11. Examine slide 54 labeled *Claviceps purpurea* sclerotia c.s. and slide 55 labeled *Claviceps purpurea* sclerotia l.s. Identify the **sclerotium** and **conidia**. To which Class does *Claviceps purpurea* belong? How does this compare to *Venturia* (slide 20)?

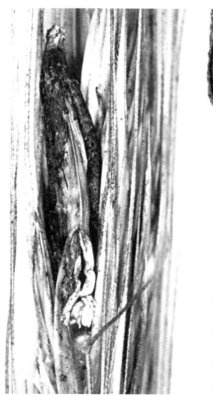
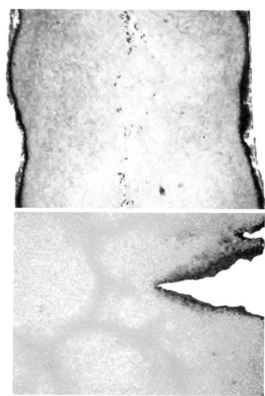

Readings

Barnett, H.L. 1960. Illustrated genera of imperfect fungi, 2nd ed. Burgess Publishing Company, Minneapolis, MN USA.

Hughes, S.J. 1953. Conidiophores, conidia and classification. Canadian Journal of Botany 31: 577-659.

Kendrick, W.B. & Di Cosmo, F. 1979. Teleomorph anamorph connections in Ascomycetes. *In*: Kendrick, W.B. (ed.) The whole fungus. Vol 1: 283-410. National Museum of Natural Sciences, Canada.

Pitt, J.I. & Samson, R.A. 2007. Nomenclatural considerations in naming species of *Aspergillus* and its teleomorphs. Studies in Mycology 59: 67-70.

Seifert, K.A. & Samuels, G.J. 2000. How should we look at anamorphs? Studies in Mycology 45: 5-18.

Sutton, B.C. 1980. The Coelomycetes. Commonwealth Mycological Institute, Kew.

Sutton, B.C. & Hennebert, G.L. 1994. Interconnections amongst anamorphs and their possible contribution to Ascomycete systematics. pp. 77-100 *In*: Hawksworth, D.L. (ed.), Ascomycete systematics: problems and perspectives in the Nineties. Plenum Press, London.

Domain Eukarya
  Supergroup - a composite organism, therefore it belongs to more than one
    Supergroup!
    Phylum Mycophycophyta (Lichens)
     Class Lichenes

Related Terms: dual organism (mycobiont+phycobiont), crustose, foliose, fruticose, hymenium, upper cortex, algal symbiont, medulla, lower cortex

Lichens

Phylum Mycophycophyta

The Mycophycophyta or the lichens are represented by nearly 25,000 species that typically occur in extreme terrestrial environments (exposed rocky outcrops, trunks and branches of trees, etc.). It was the Swiss plant anatomist/physiologist Simon Schwendener who in 1867 proposed that the lichens were a dual organism consisting of a fungus and an alga. The fungus (mycobiont) and bacteria or alga (photobiont) is how we refer to these components today. The mycobiont is typically a species of ascomycete, and rarely a species basidiomycete. Globally more than 90% of the mycobionts are ascomycetes and about 40% of the ascomycetes form lichen associations. North American lichens only have ascomycete mycobionts. The few lichen species consisting of the basidiomycete mycobiont are restricted to the tropics. The photobiont can be a species of cyanobacteria, a green alga, or yellow-green alga. The photobionts are represented by nearly 100 species, 90% of which are green alga. In this dual organism the mycobiont provides the structure within which the photobiont resides and avoids desiccation (drying up), while the photobiont provides the mycobiont with photosynthates to sustain it. The symbiotic relationship that is established between the two organisms is reported by some to represent an extreme case of mutualism while others suggest it is a parasitic or commensalistic relationship. Both mycobionts and photobionts are capable of existing independently in nature. As a dual organism lichens reproduce asexually by fragmentation, specifically through the formation of isidia (finger-like projections of the thallus cortex), or soredia (pustules of hyphae and algae/cyanobacteria on the thallus cortex).

1. Examine the lichen samples on demonstration. You must be able to distinguish among the main growth forms (crustose, foliose, fruticose, leprose, and squamulose). What characteristics did you use to make this differentiation? You must also be able to identify asexual structures (isidia and soredia) and a hymenial layer if present. Which group of fungi form the thalli in the lichens on demonstration.

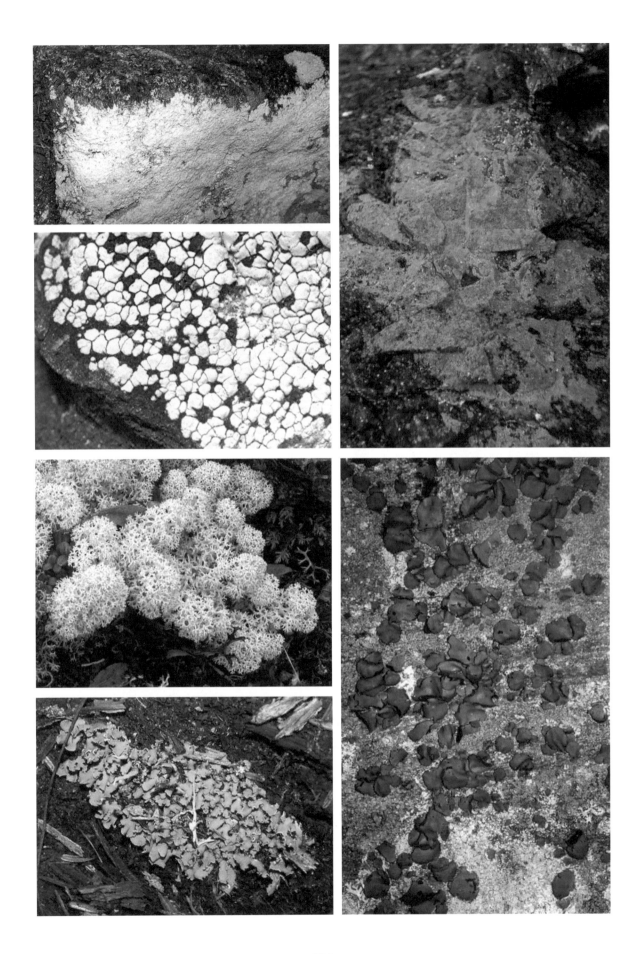

2. Examine the lichen samples on demonstration for **isidia** (papillate, bump-like projections protruding from the thallus) and **soridia** (clumps of hyphae and algae erupting from the thallus). What is/are their function(s)? — generate clones of the lichen

3. Examine the lichen samples on demonstration for **marginal cilia** (are a diagnostic characteristic for some species).

4. Observe the exposed hymenial layer of the mycobiont in the lichens on demonstration. How does the lichen (holistically) undergo sexual reproduction? Explain.

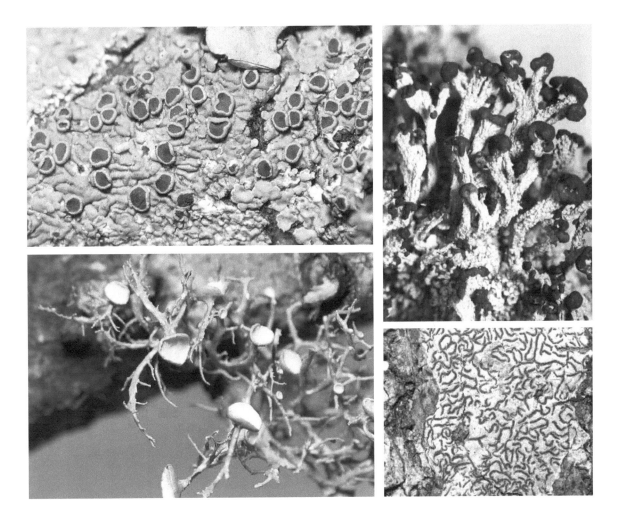

5. Examine slide 56 labeled *Umbilicaria pustulata* thallus. This is a section through a lichen thallus. What type of lichen is *Umbilicaria pustulata* (i.e. the growth form)? Label the **lower cortex**, **medulla**, **upper cortex**, **hymenium**, and **algal layer**. What type of ascoma does this lichen produce?

Crustose?
Leprose?

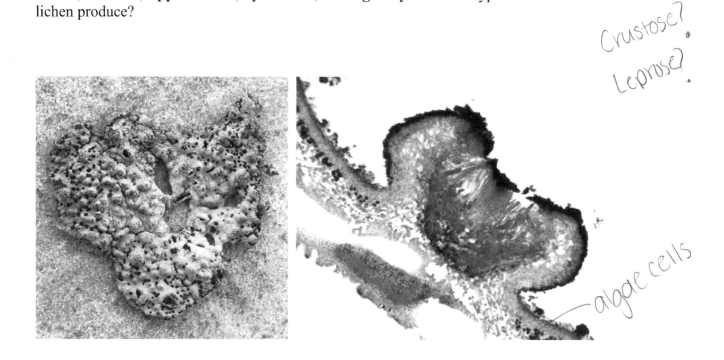

algae cells

6. Examine slide 58 labeled lichen ascocarp. Identify the **upper cortex, algal layer, medulla, lower cortex, hymenium, asci,** and **ascospores.** Are the ascospores of lichens solely responsible for sexual reproduction? Explain.

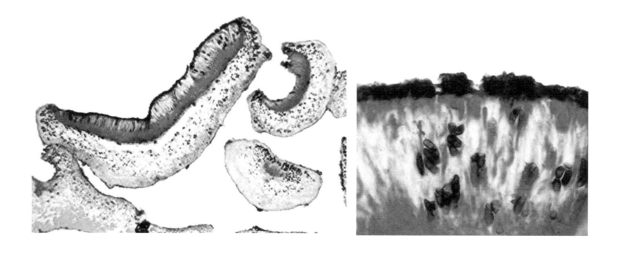

7. Examine the lichen samples on demonstration for **rhizines** (are a diagnostic characteristic for some species). Foliose

dog tooth lichen

Readings

Brodo, I.M., Sharnoff, S.D., & Sharnoff, S. 2001. Lichens of North America. Yale University Press, New Haven, Connecticut.

Hale, M.E. 1961. Lichen Handbook. A guide to the lichens of eastern North America. Smithsonian Institution, Washington, D.C.,

Hale, M.E. 1967. The biology of lichens. Edward Arnold (Publishers) Ltd., London.

Hinds, J.W. & Hinds, P.L. 2007. The macrolichens of New England: descriptions, color illustrations, identification keys, conservation status. New York Botanical Garden Press, New York.

Honegger, R. 2000. Simon Schwendener (1829-1919) and the dual hypothesis of lichens. The Bryologist 103: 307-313.

Nash, T.H. 2008. Lichen Biology 2nd ed. Cambridge University Press, Cambridge.

Sanders, W.B. 2001. Lichens: The Interface between Mycology and Plant Morphology. BioScience 51:1025-1036.

Taylor, T.N., Hass, H., Remy, W., & Kerp, H. 1995. The oldest fossil lichen. Nature 378: 244–244.

Domain Bacteria
    Supergroup - are not recognized in this Domain
        Phylum Cyanobacteria (Blue-green Algae)

Related Terms: chlorophyll *a*, carotenoids, phycoerythrin, phycocyanin, cyanophycean starch, no flagella, cellulose, pectin, oftentimes with mucilaginous sheath, sometimes calcified, 3.5 billion years, stromatolites, akinetes, heterocysts, hormogonia, mostly fresh water, only asexual reproduction

Algae

The modern definition of algae is restricted to protists (eukaryotes) that carry out the process of photosynthesis within membrane bound organelles called chloroplasts. Historically the prokaryotic Cyanobacteria which conduct photosynthesis on specialized infolded cytoplasmic membranes called thylakoid membranes were included among the algae. Algae commonly occur in aquatic ecosystems whether freshwater or marine. Less typically do they occur in terrestrial environments that include the surface of animals, plants, rocks (on and within), and soil, but also extreme environments like on snow and ice and around hot springs. In other cases we see algae growing within other organisms, namely fungi and invertebrates. The fossil record of eukaryotic algae dates back to more than 1.2 billion years. The algae are represented by more than 325,000 species with the bulk of the taxa occurring in marine systems. The algae in all likelihood represent a polyphyletic group that includes many groups depending on which classification scheme is utilized. Separation of these is dependent upon cell wall composition, photosynthetic reserve product, flagella morphology, chloroplast anatomy, photosynthetic pigments, derivation of the chloroplast and molecular data. We will restrict our examination to the brown, golden, green, red, and yellow-green algae, as well as, diatoms, dinoflagellates, euglenids, and cyanobacteria. Using today's definition, the algae are unlike the land plants in that they lack leaves, rhizoids and roots, and multicellular sex organs covered by sterile cells (there are exceptions however). Generally speaking the algae differ from the land plants (specifically the vascular plants) in their overall organization in that they lack tissue differentiation. The multicellular red algae, for example, are composed of aggregations of filaments and hence are not parenchymatous (tissue forming). As a group the algae exhibit considerable morphological variation in that they may be unicellular and microscopic, colonial, filamentous, or multicellular and macroscopic. Reproduction may be, depending on the group, asexual, sexual, or both.

Phylum Cyanobacteria

The blue-green "algae" or Cyanobacteria are prokaryotes that, according to fossil evidence originated about 3.5 billion years ago. These fossils are found within structures called

stromatolites. Living stromatolites which are layered deposits of calcium carbonate can still be found today along the coastlines of western Australia. It is believed that the blue green algal ancestors that gave rise to these stromatolites are what transformed our planet's atmosphere to one that is now oxygen rich. The ability to photosynthesize and produce oxygen as a by-product dates back 2.3 to 2.4 billion years ago. Over approximately a 2 billion year span the blue green algae were responsible for raising global atmospheric oxygen levels from 1 to 20%. Today this predominantly freshwater group is represented by nearly 2000 species. Arguably some of the more important of these are the marine picoplankton (e.g. *Prochlorococcus* and *Synechococcus*) which measure 0.2-2 μm in diameter, contribute significantly to the phytoplankton biomass (100 million + cells per liter) and account for half of the oxygen generated and carbon fixed in the world's oceans. Like other algae, blue green algae possess thylakoids, but lack chloroplasts. Furthermore, it was an endosymbiotic event in which an ancestral cyanobacterium was engulfed by a heterotrophic eukaryote that gave rise to the modern day chloroplast. This chloroplast is found in members of the supergroup Archaeplastida (containing the red and green algae). Species of blue green algae are characteristically blue-green in color due to the pigment phycocyanin, the primary pigment is chlorophyll *a*, the food reserve is cyanophyte starch, and cell walls are primarily composed of peptidoglycans. The cyanobacteria are single celled or are colonial (e.g. filaments or sheets). These bacteria are an environmentally important group as they are able to fix atmospheric nitrogen in heterocysts.

Terms:

> **heterocyst** - a differentiated cell that is responsible for nitrogen fixation.
>
> **akinete** - a thick walled dormant cell; a resting stage.
>
> **hormogonia** - part of a colony of cyanobacterial cells in a filamentous arrangement that separate from the main colony.
>
> **sheath** - a mucilaginous structure that encases cyanobacteria.

1. Examine slide 61 labeled *Gloeocapsa* w.m. The walls of blue-green algae have two layers, an inner cellulosic layer and an outer mucilaginous or pectic layer. *Gloeocapsa* is a colonial blue-green alga. *Gloeocapsa* is the simplest of all colonies, resulting when cells remain attached to each other after division. Does your roof look like this?

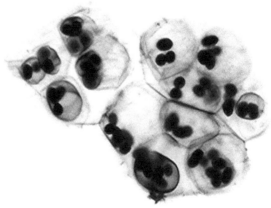

2. Examine slide 62 labeled *Lyngbya* w.m., another of the blue-green algae. *Lyngbya* is a filamentous blue-green alga which has thick sheaths. Identify the **mucilaginous sheath**, **hormogonia**, **akinetes**, and **heterocysts**.

3. Examine slide 64 labeled *Oscillatoria* w.m., another of the filamentous blue-green alga. Identify the **mucilaginous sheath**, **hormogonia**, **akinetes**, and **heterocysts**.

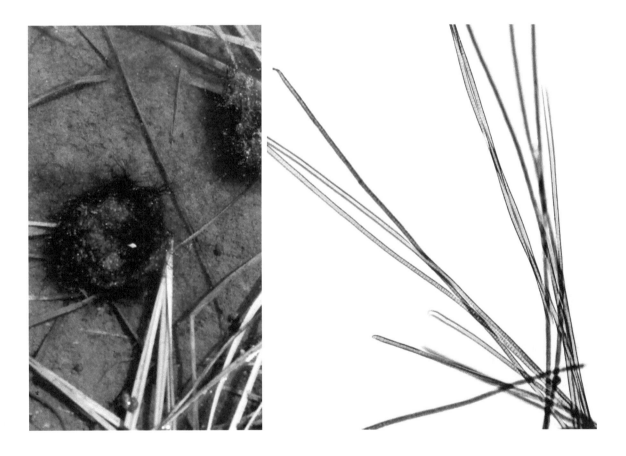

4. Examine slide 63 labeled *Nostoc* section, another of the filamentous blue-green algae. Identify the **mucilaginous sheath**, **hormogonia**, **akinetes**, and **heterocysts**. The mucilaginous sheath, an extensive gelatinous matrix in which the alga is embedded, was secreted by the filament. Heterocysts are large, clear cells and are the sites of nitrogen fixation. Akinetes are large, rounded, thick-walled, spore-like resting cells.

5. Examine the collection of dark bluish-green *Nostoc* balls. The ball represents a colony of thousands of interwoven chains of cells. Under drought conditions the ball may dry to the point of crumbling. With the addition of water some cells will begin to grow.

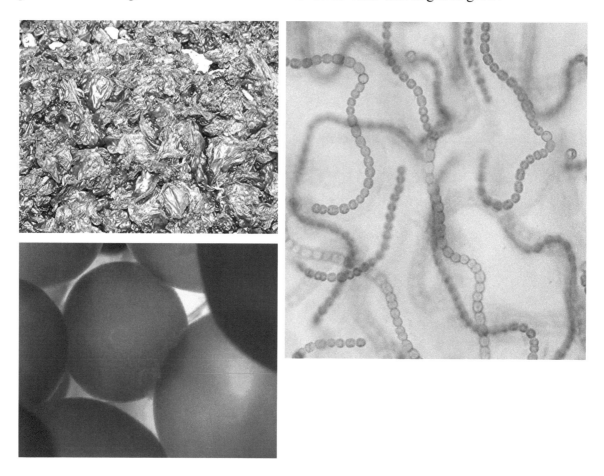

6. Examine slide 60 labeled *Anabaena* w.m., which is another of the filamentous blue-green algae. Identify the **mucilaginous sheath, hormogonia, akinetes,** and **heterocysts**. Compare *Nostoc* and *Anabaena* morphology. The genus *Anabaena* commonly clogs filters in water systems and is known to water engineers as "Annie."

7. Examine slide 65 labeled *Rivularia* w.m., another of the filamentous blue-green algae. The filaments are markedly tapered and either branched or falsely branched and aggregated into gelatinous balls. Heterocysts are usually basal and the akinetes, if present, are adjacent to the basal heterocysts.

8. Examine slide 68 labeled *Spirulina*, a filamentous blue-green algae. *Spirulina* is abundant in Na_2CO_3 lakes north of Lake Chad and in East Africa. Chemical analysis of algal cakes sold at local markets showed 45-49% protein by dry weight, suggesting that *Spirulina* may be a promising local food source. Compare the morphology with *Oscillatoria*.

9. Examine slide 66 labeled *Aphanizomenon*, a filamentous blue-green algae. The genus *Aphanizomenon* commonly clogs filters in water systems and is known to water engineers as "Fanny." Draw *Aphanizomenon* i.e. "Fanny" and compare to *Anabaena* i.e. "Annie". Identify differences between "Fanny" and "Annie".

10. Examine slide 67 labeled *Merismopedia*, a colonial, sheath like blue-green algae. Note the sheath.

11. Prepare a wet mount of the blue-green algae growing on one of the pebbles under a bench in the greenhouse. Watch the slow gliding and waving movements of the living filaments of *Oscillatoria* and related genera. Identify the **mucilaginous sheath**, **hormogonia**, **akinetes**, and **heterocysts**. Illustrate and label what you see. Additionally, examine all living cultures of cyanobacteria on demonstration.

Readings

Bhattacharya, D., & Medlin, L. 1998. Algal phylogeny and the origin of land plants. Plant Physiology 116: 9-15.

Fogg, G.E., Stewart, W.D.P., Fay, P., & Walsby, A.E. 1973. The Blue-green Algae. Academic Press, NY.

Whitton, B.A. & Potts, M. 2000. The Ecology of Cyanobacteria: Their diversity in time and space. Kluwer Academic Publishers, Netherlands.

Domain Eukarya
Supergroup Excavata
Phylum Euglenophyta (Euglenoids)

Related Terms: chlorophyll *a* and *b*, carotenoids, xanthophylls, paramylon granules, 2 apical flagella, no cell wall (pellicle), gullet, ampulla, facultative autotroph, mostly fresh water and pollution tolerant, non-planktonic, phagocytosis, serial endosymbiosis, euglenoid movement, only asexual reproduction

Euglenids

Phylum Euglenophyta

The Euglenophyta are represented by nearly 1,000 species that most commonly occur in freshwater systems and far less commonly in marine environments (pools, ponds, lagoons, and bays). In both systems an abundance of decaying organic matter is required. Euglenid species are both pollution and pH tolerant to acid environments (to a pH of 2). Species are unicellular or colonial, most are autotrophic, some are mixotrophic (they can exist as autotrophs or heterotrophs). However, botanists typically treat them as an alga as most are autotrophic. Euglenids may possess the pigments chlorophyll *a* and *b* as well as several xanthophylls which result in a green pigmentation like that of the green algae. However, the Euglenophyta differ from the green algae in their molecular content, cellular organization and biochemistry. Further, plastids are believed to be derived by a secondary endosymbiotic event of an ancestral green alga. Euglenids lack a cell wall, but do possess a spiraled proteinaceous layer just beneath the plasma membrane known as the pellicle which is in part responsible for the organism's characteristic movement or metaboly (euglenoid movement). Additionally, euglenids may be propelled in a cork screw-like motion by their 1-2 emergent flagella. The heterotrophic euglenids ingest food/prey items like bacteria by phagocytosis. Euglenids are able to sense light in part through a structure that is referred to as an eyespot (this structure indirectly aids the organism in orienting itself toward that light). All store fuel-molecules as paramylon (not starch). Because euglenids share characteristics of both plants (photosynthesize) and animals (heterotrophic, movement, and phagocytosis), their taxonomic placement throughout time has varied, thus their current placement in Kingdom Protista.

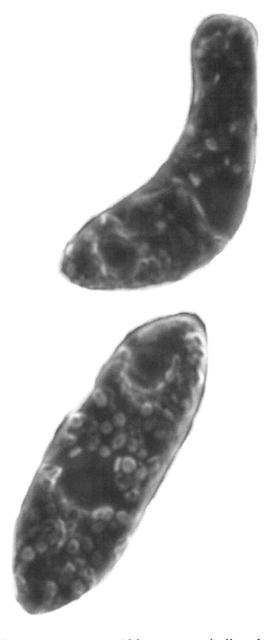

1. Examine slide 69 labeled *Euglena*. Identify the **gullet**, **eyespot**, **flagellum**, **paramylon bodies**, **chloroplasts**, **vacuole**, and **nucleus**.

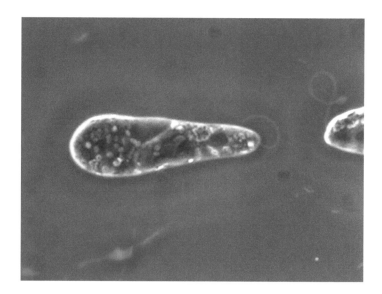

2. Examine the living culture of *Euglena*. Describe euglenoid movement.

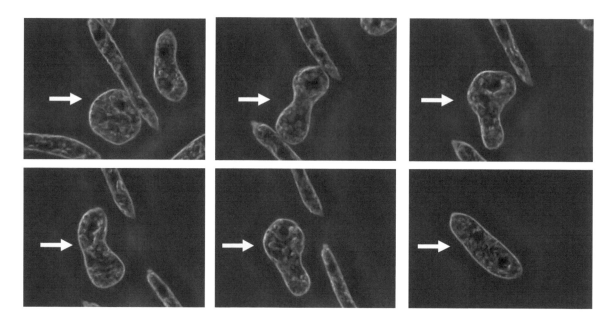

Readings

Olaveson, M.M. & Stokes, P.M. 1989. Responses to the acidophilic alga *Euglena mutablis* (Euglenophyceae) to carbon enrichment at pH 3. Journal of Phycology 25 529-539.

Olaveson, M.M. & Nalewajiko, C. 2000. Effects of acidity on the growth of two *Euglena* species. Hydrobiologia 433 39-56.

Wehr, J.D. & Sheath, R.G. 2003. Freshwater algae of North America: ecology and classification. Academic Press, San Diego, CA.

Domain Eukarya
    Supergroup Chromalveolata
        Phylum Dinophyta (Dinoflagellates)   *Peridium, Gonyaulax, Ceratium*

Related Terms: chlorphylls *a* and *c*, carotenoids, zanthophylls, starch, lipids, 1 girdle, 1 sulcus, cellulose pectin, mostly marine, thecal plates, saxitoxin, paralytic shellfish poisoning, PSP, red tide, filter feeding shellfish, fish kills, asexual reproduction is predominant, halves break and replace, sexual reproduction rarely occurs, zygote, cyst, resting stage

Dinoflagellates

Phylum: Dinophyta

The Dinophyta (Pyrrophyta) are represented by approximately 2,000 freshwater and marine systems species. A small number of species have been reported from wet snow or sand. Dinoflagellates are commonly unicellular, though some genera are multicellular filaments. Like the euglenids, some are autotrophic, some are heterotrophic, and some are mixotrophic. The autotrophic species are only second in importance to the diatoms as primary producers in coastal marine waters. The Dinophyta are unusual algae in that some have undergone secondary endosymbiosis, whereas others have undergone tertiary endosymbiosis. Those species that have secondarily lost their chloroplasts are more animal-like than plant-like. The photosynthetic species contain chlorophyll *a* and *c* with the carotenoids mainly being the accessory pigment peridinin. Some

photosynthetic species are endosymbionts called zooxanthellae and are important in coral reef structure. These zooxanthellae also occur in other invertebrate groups like sea anemones, jellyfish, flatworms, clams, and protozoans. As coral endosymbionts the dinoflagellate increases the coral's chance for survival in nutrient poor waters. The external structure of dinoflagellates is distinct and can only be appreciated through electron microscopy imagery. The body, which is referred to as a theca, is covered in cellulosic plates which may bear spines or horns. A central horizontal groove known as the girdle and a vertical groove known as the sulcus each possess a single flagellum. Most dinoflagellates are biflagellate. The flagellum associated with the girdle is used for rotation/orientation and that associated with the sulcus is used for propulsion. Contrary to their significant role in carbon cycling, dinoflagellates may negatively impact the environment through the production of blooms (concentrations of cells in excess of 1,000-20,000 per ml) which may be toxic to various aquatic species. These blooms are typically referred to as red tides because they color the water reddish. The toxins produced by the dinoflagellates may bio-accumulate in those organisms that have directly ingested them. These organisms may in turn be toxic to those organisms that consume them. The symptoms related to the consumption of these toxins are oftentimes associated with the gastrointestinal, nervous, or cardiac systems. Death may result. Consequently, these dinoflagellate blooms are of particular interest to the commercial fishing industry. Other dinoflagellates are interesting for their entertainment value as they bioluminesce in the blue-green wavelengths. Another characteristic of the dinoflagellates that is not common among other organisms is its permanently condensed chromosomes. Reproduction in dinoflagellates may be sexual or asexual. Asexual reproduction involves little more than the cell splitting along the girdle and subsequent generation of the missing half. Sexual reproduction involves the differentiation of vegetative cells into gametes followed by their fusion which results in the formation of the diploid planozygote, followed by the resting hypnozygote, and the emergence of the haploid planomeiocyte (i.e. a vegetative cell) following meiosis. The dinoflagellates life cycle is an example of zygotic meiosis.

1. Examine slide 77 labeled *Ceratium* w.m. Identify the **thecal plates**, **girdle (cingulum)**, **sulcus**, and **flagella** if possible. Species like *Ceratium* have horn-like projections. What function do these serve? Note the thecal plates from dinoflagellates that have disintegrated.

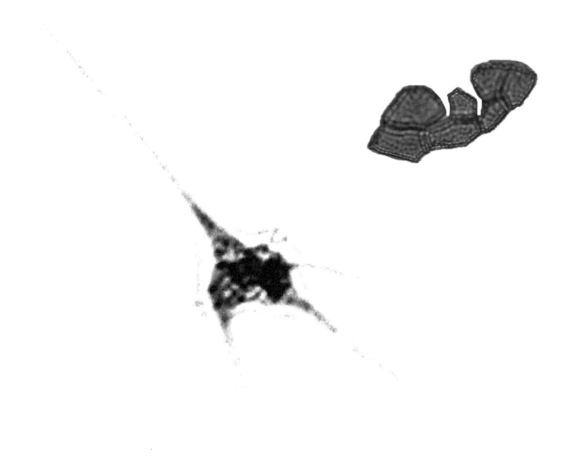

Readings

Burkholder, J.M., Noga, E.J., Hobbs, C.W., Glasgow, H.B. Jr., & Smith, S.A. 1992. New "phantom" dinoflagellate is the causative agent of major estuarine fish kills. Nature 358: 407-410.

Faust, M.A. & Gulledge, R.A. 2002. Identifying harmful marine Dinoflagellates. Contributions from the United States National Herbarium 42. Washington, DC: Department of Systematic Biology, Botany, National Museum of Natural History.

Hackett, J. D., Anderson, D., Erdner, D., & Bhattacharya, D. 2004. Dinoflagellates: A remarkable evolutionary experiment. American Journal of Botany 91: 1523-1534.

Morris, J.G. Jr. 2001. Human Health Effects and *Pfiesteria* Exposure: A Synthesis of Available Clinical Data. Environmental Health Perspectives 109: 787-790.

Steidinger, K.A. & Tangen, K. 1997. Chapter 3, Dinoflagellates. *In*: Tomas, C.R. (ed.), pp. 378-584. Identifying Marine Phytoplankton. Academic Press, San Diego.

Taylor, F.J.R. (ed.). 1987. The Biology of Dinoflagellates. Botanical Monographs 21: 1-785.

Domain Eukarya
      Supergroup Chromalveolata
            Phylum Oomycota (Water Molds)
                Order Saprolegniales
                Order Peronosporales

Related Terms: oogonia, antheridia, female gametes, biflagellate zoospore, oogonia with 1 oospore

Domain Eukarya
      Supergroup Chromalveolata
            Phylum Heterokontophyta (Heterokonts)
                Class Bacillariophyceae (Diatoms)
                    Order Centrales (Centric Diatoms)
                    Order Pennales (Pennate Diatoms)

Related Terms: chlorophyll *a* and *c*, carotenoids, fucoxanthin, chrysolaminarin, 1 tinsel in sperm, silica, pectin, marine and fresh water, frustule, valves, epitheca, hypotheca, girdle, raphe, striae, punctae, asexual reproduction, shrinking diatom syndrome, sexual reproduction, anisogamy, oogamy, fusion of gametes, zygote, centric, pennate, diatomaceous earth, reproductive/evolutionary dichotomy!

Class Chrysophyceae (Golden Algae) *Synura, Dinobryon*

Related Terms: chlorophyll *a* and *c*, carotenoids, fucoxanthin, chrysolaminarin, 1 tinsel, 1 whiplash, silica, pectin, mostly fresh water, unicellular, colonial, lorica, fishy smelling ketones and aldehydes, asexual reproduction (divide in unicellular form, fragment in multicellular form), sexual reproduction (isogamy)

Class Xanthophyceae (Yellow-green Algae)

Related Terms: chlorophyll *a* and *c* (*c* in low quantities), carotenoids, vaucheriaxanthin, chrysolaminarin, 1 tinsel, 1 whiplash, cellulose, silica in some, mostly fresh water

Class Phaeophyceae (Brown Algae)
      Order Ectocarpales   *Ectocarpus*
      Order Dictyotales   *Dictyota, Padina*
      Order Laminariales   *Postelsia, Alaria, Macrocystis, Laminaria, Agarum, Nereocystis*
      Order Fucales   *Fucus, Sargassum, Ascophyllum, Hormosira*

Related Terms: chlorophyll *a* and *c*, carotenoids, xanthophylls, laminarin, 2 lateral, 1 whiplash, 1 tinsel, cellulose, pectin, alginic acids, almost all marine, isogeneratae, heterogeneratae, cyclosporae, meiospore, gametothallus, plurilocular gametangium, isogametes, isogamy, zygote, plurilocular sporangium, biflagellate zoospore, mitospore, asexual reproduction, unilocular meiosporangium, meiosis, anisogamous-oogamous, kelps, seaweeds, holdfast, blade, lamina, stipe, intertidal zone, float baldder, alginic acid, meiosporangium, receptacles, conceptacles, paraphyses, antheridium (64 spermatia), oogonium (8 eggs), chemical attractant, alginic acid, fireproofing, laundry starch, lipstick, stabilize foods, etc.

Stramenopiles

Stramenopiles or Heterokonts include: (1) the commonly parasitic oomycetes, (2) the generally unicellular diatoms, (3) the unicellular or colonial golden algae, (4) the multicellular brown algae (including the giant Kelps), (5) the unicellular, multicellular filamentous, and colonial yellow-green algae (Fig. 1). Although both autotrophic and heterotrophic organisms are included in the group they are unified by their collective possession of heterokont flagella (flagellate cells possessing two anatomically different flagella) at some stage in their life cycle as well as the acquisition of a chloroplast through a secondary endosymbiotic event most likely from an ancestral red algal unicell. As some groups are heterotrophic this chloroplast has been secondarily lost. Molecular data further supports the monophyly of this group. The algal taxa are further unified by the possession of chlorophyll c thus their inclusion in the chromophytes (i.e. algae that contains chlorophyll c).

Figure 1. Cladogram of the evolutionary relationships of the oomycetes, diatoms, golden algae, brown algae, and yellow-green algae.

Common names for Stramenopile lineages:

brown algae – Class Phaeophyceae
diatoms – Class Bacillariophyceae
golden algae – Class Chrysophyceae
oomycetes – Phylum Oomycota
    downy mildews – Order Peronosporales
    water molds – Order Saprolegniales
yellow-green algae – Class Xanthophyceae

Lect.3

Phylum Oomycota

The Oomycota is a group of fungal-like protists, which are believed to have evolved from a shared common ancestry with chromophyte algae (specifically the Bacillariophyceae, Chrysophyceae, Phaeophyceae, Xanthophyceae) rather than other groups of fungi. The Oomycota share many features in common with these algae including life cycles that undergo gametic meiosis. Gametes are oogamous. The Oomycota is a heterotrophic group with many of its members being parasitic on animals and plants. The phylum includes nearly 700 species which are either aquatic (Saprolegniales) or terrestrial (Peronosporales). The terrestrial species are dependent on the availability of

large amounts of water for their continued survival, as specific reproductive cells are flagellated. The importance of this group rests in its propensity to attack various crop plants and has even resulted in the emigration of people from those areas as the destruction of crops has led to starvation in specific regions of the world. One species, *Phytophthora infestans* was responsible for the the Great Irish Potato Famine of the 1840s that resulted in nearly one million deaths due to starvation. In France during the 1870s the Oomycete *Plasmopara viticola* nearly destroyed the wine industry; however a chemical treatment was devised to treat the disease, namely Bordeaux mixture.

Order Saprolegniales

1. Examine slide 3 labeled *Saprolegnia* reproductive which is a water mold (Oomycota, Saprolegniales). Label the sexual reproductive structures **oogonial initial, oogonium, eggs,** and **antheridia**, as well as **the sterile hyphae**. You may see a fertilization tube and the fertilized eggs. Where are the flagella? Diagram and label the life cycle of *Saprolegnia* in the space below.

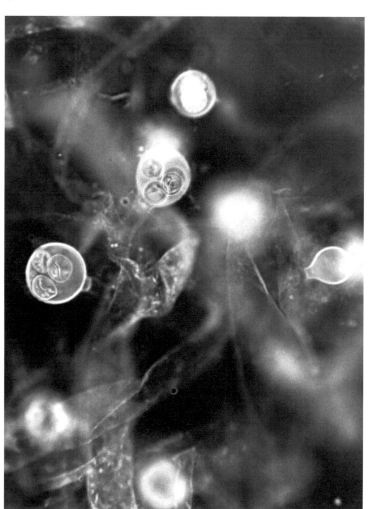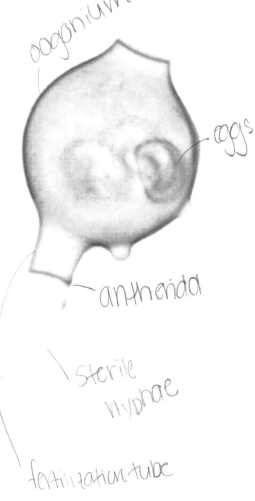

2. Examine the living culture of *Saprolegnia*. Make a slide preparation. Observe and identify **zoosporangia (asexual)** and **oogonia (sexual)** structures if present.

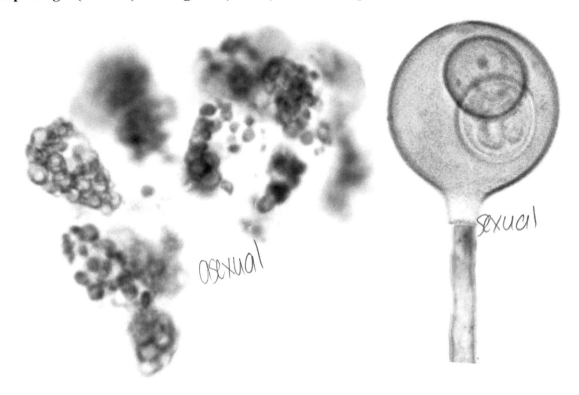

asexual

sexual

Order Peronosporales   downy mildews

3. On demonstration are leaves and berries infected with "downy mildew" of grape which is caused by the fungal-like protist *Plasmopara viticola*.

128

4. Examine slide 4 labeled *Plasmopara viticola* stem c.s. and note the branched aerial sporangiophores and deciduous sporangia. How is this fungal-like protist <u>dispersed?</u>

wind

—sporangium

—sporangiasp[ore]

—sporangium

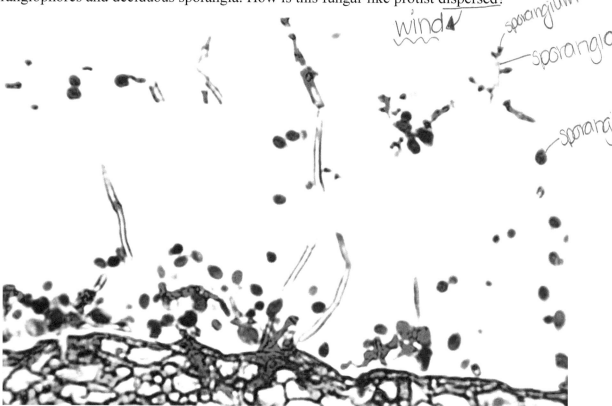

5. Examine slide 5 labeled *Plasmopara viticola* leaf c.s. and note the branched aerial sporangiophores and deciduous sporangia. How is this fungus-like protist dispersed? Which is the adaxial surface of the leaf? Explain.

wind

axidal surface

sporangiospore

sporania

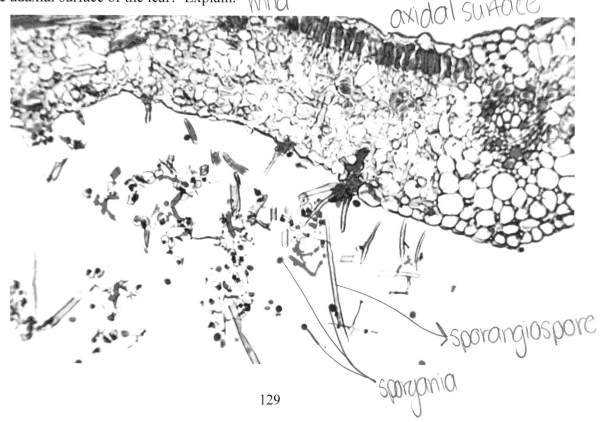

129

6. Examine slide 7 labeled *Phytophthora infestans* stem and note the branched aerial sporangiophores and deciduous sporangia. How is this fungal-like protist dispersed? wind?

7. Examine slide 10 labeled *Phytophthora infestans* tuber and note the hyphae.

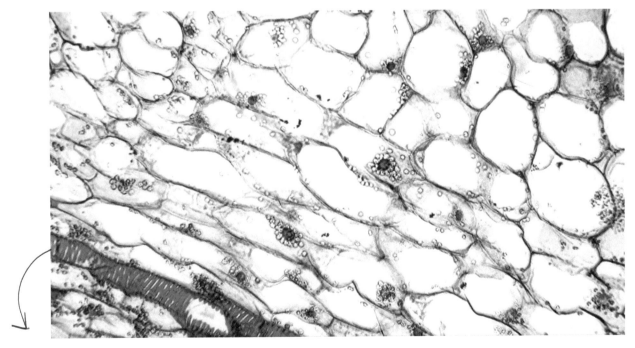

hyphae

8. Examine slide 11 labeled *Phytophthora infestans* leaf and note the branched aerial sporangiophores and deciduous sporangia. How is this fungus-like protist dispersed?

rain/wind

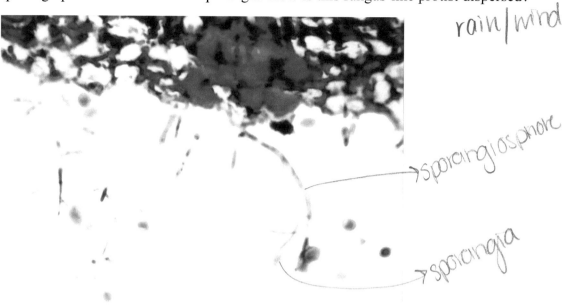

→ sporangiosphore

→ sporangia

Phylum
Oomycota

9. Examine slide 12 labeled *Peronospora parasitica* and indentify the **sporangiophores** and **sporangia**. Are the sporangiophores **haploid** or diploid? Are the sporangia **mitosporangia** or **meiosporangia**? Are the spores generated from the sporangia zoospores or **non-flagellated spores**?

branch
that bec

mitosis
mitosporangium

sporangia → zoospores
2n

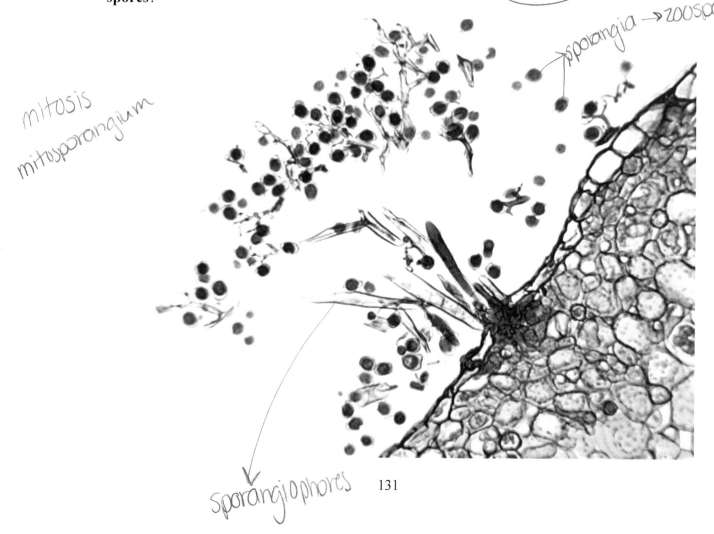

sporangiophores

131

Phylum: Heterokontophyta

The Heterokontophyta as a group exhibits considerable morphological variability and includes the diatoms, brown algae, golden algae, and yellow-green algae. Species within this group possess both tinsel and whiplash flagella. Chloroplasts lie up against the nucleus and are enclosed by two membranes of the endoplasmic reticulum. Thylakoids usually occur in stacks of three. Pigments present include chlorophyll *a* and *c* (chlorophyll *b* <u>never</u> present) with fucoxanthin as the primary accessory pigment (except in Xanthophyceae, that contain vaucheriaxanthin). Reserve photosynthate that is formed in vacuoles outside the chloroplast is stored as chrysolaminarin.

Class: Bacilliariophyceae

Bacilliariophyceae represent a very large group of unicellular marine, freshwater, terrestrial and sub-aerial algae. Approximately 10,000-12,000 species are currently recognized with an estimated diversity of 100,000-10,000,000 species. As primary producers, diatoms are the foundation of the oceanic food chains and in that regard they are ecologically important. Furthermore, marine diatoms, due to their abundance and productivity, are responsible for approximately 25% of the world's carbon fixation contribution. Their commercial significance rests in the production of bio-fuels, and their use as abrasives and filtering aids (diatomaceous earth). Some species (e.g. *Pseudo-nitzschia pseudodelicatissima*) contain toxins, namely domoic acid, which is responsible for amnesic shellfish poisoning. Diatoms may be planktonic and one of the most common types of phytoplankton or benthic (as epiphytic, epilithic, and etc.), and may be solitary or colonial. Pigments include chlorophyll *a* and *c* as well as fucoxanthin. The diatoms have traditionally been divided into two orders based predominately on their cell wall or frustule morphology. The cell walls are made of silica and can be major contributors to ocean sediments (diatomaceous earth) in deep waters. The frustule per se consists of the larger epitheca which partially overlaps the smaller hypotheca in the region of the girdle. Both halves of the frustule may be perforated by pores called punctae which may be lined in rows called striae. Simplistically two Orders are recognized, the Pennales (pennates and araphid pennates) and the Centrales (centrics). Both sexual and asexual reproduction occur.

10. Examine slide 70 labeled Diatoms w.m. Diatoms are unicellular chromophyte algae that live in silica boxes. They are abundant in both fresh and marine waters and are very important as primary producers. Can you find both centric and pennate diatoms? Explain! Can you identify the **epitheca**, **hypotheca**, **girdle**, **raphe**, **punctae**, and **striae**? To which order does each of the following frustules belong? Use your condenser and fine focus adjustments to see the fine detail.

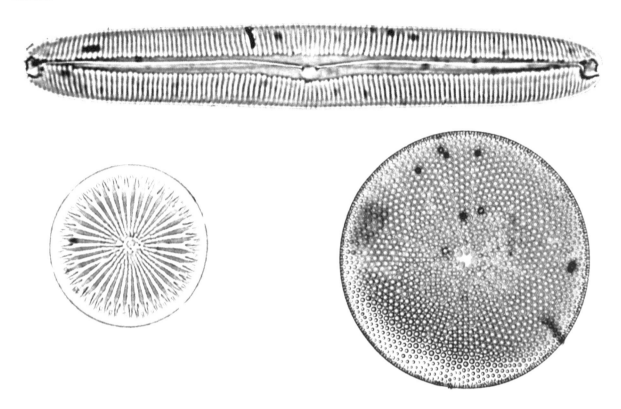

11. Examine slide 71 labeled diatomaceous earth w.m. for the various forms of diatom frustules which may be broken into fragments. Diatomite is the trade name of diatomaceous earth, a marine deposit of fossil diatoms, which is used as a filtering agent, as a mild abrasive (in toothpaste), and as a "natural" insecticide. Which are more common in this slide of diatomaceous earth the **centric** or **pennate** diatoms?

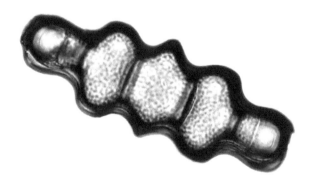
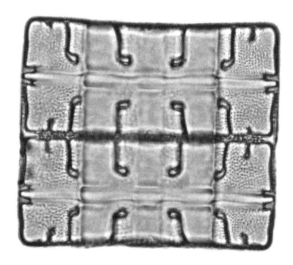

133

Class: Phaeophyceae (brown algae)

The brown algae are predominately marine with only some freshwater representatives. This class of algae consists of more than 1500 species that are divided among one of three life cycles, the isogeneratae which exhibit isomorphic alternation of generations (Order Ectocarpales), the heterogeneratae which exhibit heteromorphic alternation of generations (Order Laminariales), or the cyclosporae which do not exhibit alternation of generations (Order Fucales). The most notable of the brown algae are the kelps (Laminariales). Kelps are large, brown benthic macro-algae (the marine macro-algae are collectively referred to as seaweeds) that occur in nutrient rich marine waters with cold water currents. Kelp such as *Macrocystis* and *Nereocystis* form underwater forest communities in the sub-tidal zone to depths of

approximately 70m. These communities create habitat for a wide array of invertebrate and vertebrate life. Although members of the Fucales are intertidal some members of the genus *Sargassum* spend their entire existence free-floating in oceanic currents throughout regions of the Atlantic Ocean in the vicinity of the Tropic of Cancer, specifically the Sargasso Sea. Like kelp forests, these *Sargassum* mats create habitat for invertebrate and vertebrate organisms. The Ectocarpales typically inhabit the intertidal zone though differ from the Fucales and Laminariales in that some members lack parenchymatous construction, but instead are composed of uniseriate filaments. The brown algae are also an economically important group for iodine, soda ash (sodium carbonate = water softener), and alginate. Alginate extracts can be used as emulsifiers (thickeners) in foods (e.g. ice cream) and industrial goods (e.g. plastics, laundry starches, etc.). Some cultures even harvest specific brown algae as a food source namely, *Alaria* and *Laminaria*. The chloroplasts contain chlorophyll *a* and *c*, as well as fucoxanthin which results in their brown coloration. Four membranes surround the chloroplasts. Life cycles seen in this group include sporic and gametic meiosis.

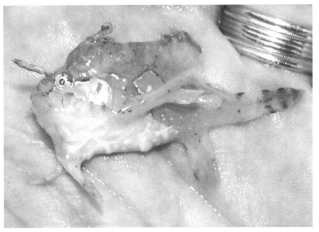

12. Examine slide 96 labeled *Ectocarpus* pluriocular w.m. Identify the **filament, plurilocular sporangium,** and **unilocular sporangium**? What is meant by the term **uniseriate** in *Ectocarpus*? Differentiate between a plurilocular **sporangium** or **gametangium**?

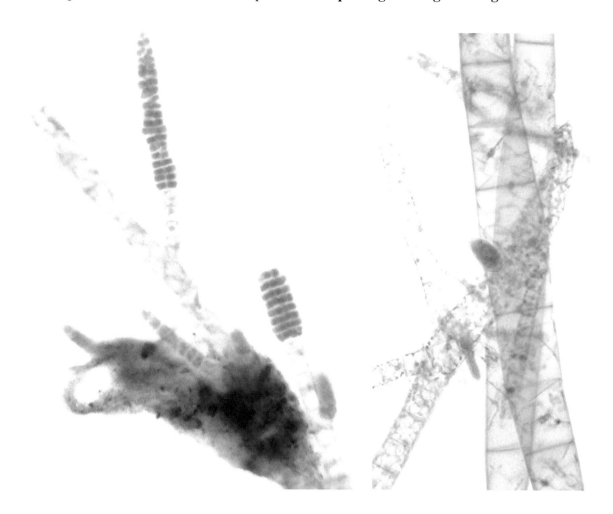

13. Examine slide 98 labeled *Macrocystis* c.s. with sieve plates. What is the function of sieve plates and where do they occur? Why are so few sieve plates visible in the field of view?

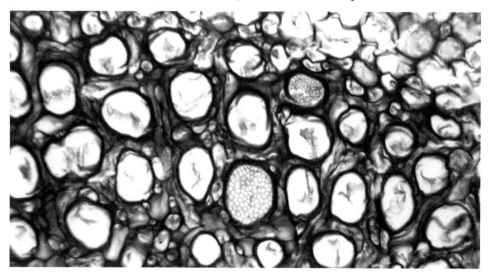

14. Examine slide 97 labeled *Laminaria* sporangia. Identify the **lamina, unilocular sporangia** and associated **paraphyses**. Where do the unilocular sporangia occur? Identify the **ploidy level** (i.e. **haploid** or **diploid**) of each structure. Where does meiosis specifically occur in this organism?

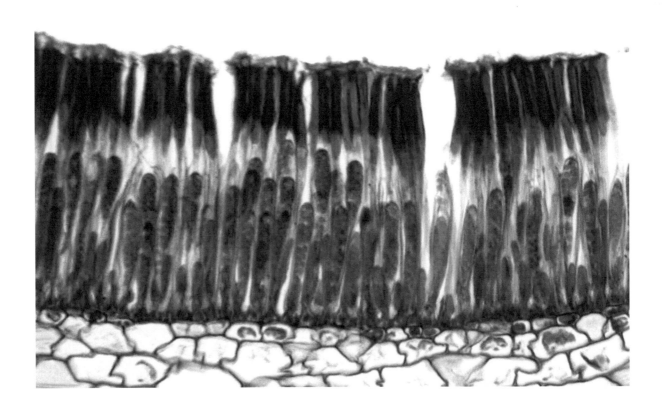

15. Draw one of the herbarium or preserved specimens of the kelp *Laminaria* and label the **holdfast, stipe, fertile region,** and **lamina**. Is this a **gametothallus** or **sporothallus**? Is the thallus **haploid** or **diploid**? What type of life cycle is exhibited by *Laminaria*? What pigment is responsible for the brown coloration of this alga?

16. Examine slide 72 labeled *Fucus* male conceptacles median l.s. In *Fucus* the gametes are formed in chambers called **conceptacles** on swollen, terminal portions of the thallus called **receptacles**. Each male gametangium yields 64 sperm. Identify the **conceptacles**, **antheridia** and **paraphyses** (have you seen this term previously?).

17. Examine slide 73 labeled *Fucus* female conceptacles median l.s. Each female gametangium yields 8 eggs. Identify the **oogonium** and **eggs**. What type of life cycle does *Fucus* exhibit? Diagram and label the life cycle of *Fucus* below.

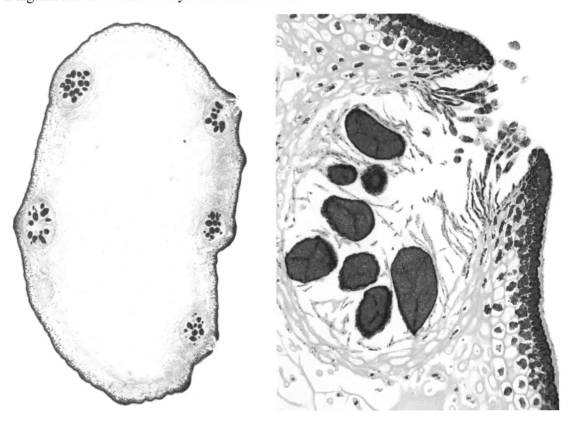

138

18. Is the pelagic *Sargassum* thallus shown below **haploid** or **diploid**? Identify the **receptacles** and **air bladders** on the thallus. What is the main function of receptacles? What other Order of brown algae has air bladders?

19. Examine slide 74 labeled *Sargassum* section. The genus *Sargassum* reproduces sexually if anchored, but asexually if free-floating. Antheridial and oogonial conceptacles are produced together on fleshy receptacles during the course of sexual reproduction. Excised segments containing apical or lateral meristems are able to continue growth in culture. Identify the location of at least one **ostiole** (have you seen this term before?) on a **conceptacle**.

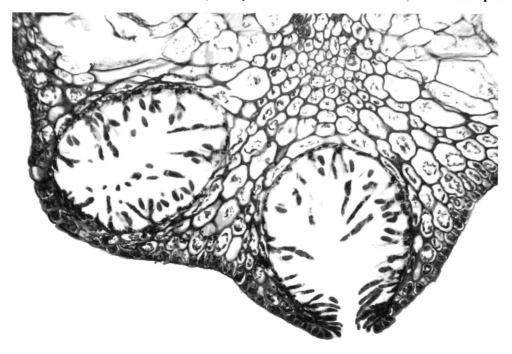

Readings

Andersen, R.A. 2004. Biology and systematics of heterokont and haptophyte algae. American Journal of Botany 91: 1508-1522.

Fredericq, S., Cho, T., Earle, S.A., Gurgel, C.F.D., Krayesky, D.M., Mateo Cid, L.E., Mendoza Gonzáles, A.C., Norris, J.N., & Suárez, A.M. 2009. Seaweeds of the Gulf of Mexico. *In*: Felder, D.L. and Camp, D.K. (eds.), pp. 187-259, Gulf of Mexico: Its Origins, Waters, and Biota, III: Biota. College Station: Texas A&M University Press.

Fry, W.E. & Grunwald, N.J. 2010. Introduction to the Oomycetes. The Plant Health Instructor 2010-1207-01 DOI: 10.1094/PHI-I-2010-1207-01.

Kingsbury, J.M. 1969. Seaweeds of Cape Cod and the Islands. The Chatham Press Inc. Chatham, MA, USA.

Krayesky, D.M., Meave del Castillo, E., Zamudio, E., Norris, J.N., & S. Fredericq. 2009. Diatoms (Bacillariophyta) of the Gulf of Mexico. *In*: Felder, D.L. and Camp, D.K. (eds.), pp. 155-186, Gulf of Mexico: Its Origins, Waters, and Biota, III: Biota. College Station: Texas A&M University Press.

Millardet, P.M.A. 1885. The discovery of Bordeaux mixture. Phytopathological Classic translated into English by F.J. Schneiderhan. American Phytopathological Society Press, St. Paul MN.

Zwankhuizen, M.J., Govers, F., & Zadoks, J.C. 1998. Development of potato late blight epidemics: Disease foci, disease gradients, and infection sources. Phytopathology 88: 754–763.

Domain Eukarya
    Supergroup Archaeplastida
        Phylum Rhodophyta (Red Algae)
            Class Bangiophyceae   *Porphyra* (Nori)
            Class Florideophyceae
                Order Ceramiales   *Polysiphonia*
                Order Corallinales   *Corallina*
                Order Gelidiales   *Chondrus* (Irish Moss)
                Order Palmariales   *Palmaria* (Dulse), *Rhodymenia*

Related Terms: chlorophyll *a* (*d*), carotenoids, phycoerythrin, phycocyanin, floridean starch, no flagella, cellulose, pectin, agar, carrageenin, sometimes calcified, mostly marine, tetraspores, male gametophyte, female gametophyte, spermatangium, spermatia, trichogyne, carpogonium, pericarp, carposporothallus, carpospore, tetrasporothallus, tetrasporangium, holdfast, unilocular sporangia, many spermatia per spermatangium, many carpospores per carposporangium versus 1, isomorphic alternation of generations, heteromorphic alternation of generations

Red Algae

Phylum Rhodophyta

The Rhodophyta are predominately a marine group with some freshwater representatives. This diverse lineage is represented by approximately 4,000-6,000 species. The red algae occur from the intertidal zone to very deep subtidal waters up to 285 m in depth. Red algae are utilized as a food source (e.g. nori) and some groups contain chemical derivatives (e.g. phycocolloids such as agar and *kappa*-carrageenans) that have nutritive, medical, scientific, pharmaceutical, or industrial value. Many of the foods we eat and products we use every day such as  toothpaste, salad dressings, detergents and ice cream to name a few contain compounds derived from red algae. The pigment systems of the red algae include chlorophyll *a* and *d* (in some groups), phycoerythrin, phycocyanin, and allophycocyanin. It is the water soluble pigment phycoerythrin that gives red algae their red coloration. Chloroplasts are derived from a primary endosymbiosis of ancestral cyanobacteria. The cell wall is primarily composed of cellulose and sulfated polysaccharides. The cells of the subclass Floridiophyceae generally undergo an incomplete cytokinesis leaving an aperture between newly formed cells that are occluded by a structure called a pit plug that separates the cytoplasm of the two cells (see below). One of the most notable characteristics of this group is its reproductive biology, namely, the triphasic life cycle exhibited by members of subclass Floridiophyceae. Spermatia (male gametes) are non-flagellated and hence unable to move on their own, but instead are subject to the vagaries of water currents. Fertilization occurs as a consequence of the chance movement of spermatia laden water around the red algal thallus in the vicinity of the trichogyne, a filamentous extension of the carpogonium (the egg cell). Although this mode of spermatia transfer seems unreliable, the subclass Floridiophyceae remains very successful due to its capitalization of the triphasic life cycle (alternation of generations). The subclass Bangiophyceae only has a biphasic life cycle (alternation of generations).

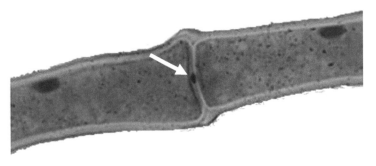

Terms:

parenchymatous - a term applied in the algae to a tissue type that is composed of parenchyma cells where there is evidence that cell division is proceeding in more than one plane.

pseudoparenchymatous - term applied in the algae to a tissue type that superficially appears parenchymatous, but is actually composed of branched filaments.

1. Examine slide 76 labeled *Polysiphonia* composite or the *Polysiphonia* preserved material.

    a. Haploid spermatia are produced within the spermatogonia on the male gametophyte. Examine the female gametophyte producing a single celled carpogonium. The carpogonium is part of a several celled structure called the carpogonial branch. It is inconspicuous and hence difficult to see.

    b. Cystocarps are formed after fusion of the male and female nuclei, and consist of a group of filaments that actively produce diploid carpospores surrounded by an urn-shaped pericarp with a conspicuous ostiole at the distal end. Carpospores develop into tetrasporophytes.

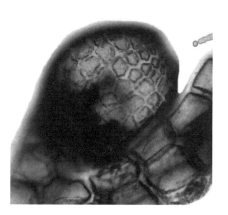

    c. Carposporophytes are produced within the pericarp. Identify the **carposporophyte** and **pericarp.** Identify which structures are **haploid** and which structures are **diploid**. What process gives rise to the first cell of the **carposporophyte (mitosis, meiosis, or syngamy)?** Explain.

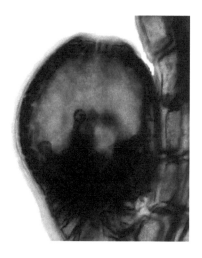

143

d. The tetrasporophyte produces tetraspores within tetrasporangia. Note the production of **tetrasporangia** and **tetraspores**. Identify which structures are **haploid** and which structures are **diploid**.

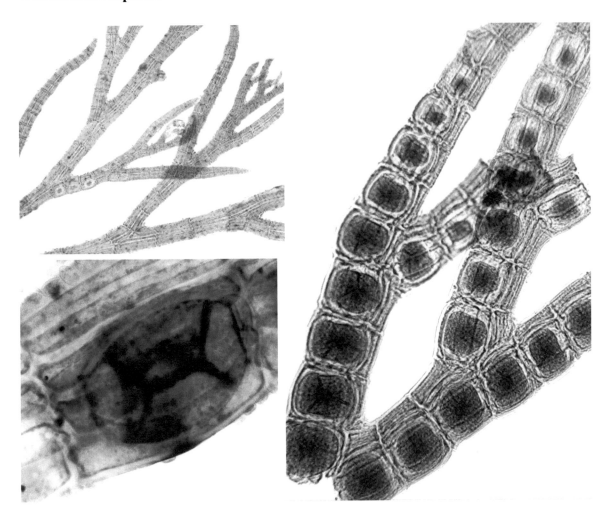

2. The red algae are not parenchymatous. Use the image of *Callithamnion biseriatium* to explain why the species is not parenchymatous?

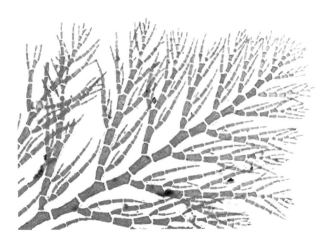

3. Below is an image of *Bostrychia radicans*. Identify the **gametophyte, pericarp, carposporophyte** and **carpospores**. The carposporophytes are completely dependent upon the female gametophyte. Identify which structures are **haploid** and which structures are **diploid**.

4. Where does meiosis occur in the life cycle of a red alga (**spermatium, carpogonium, cystocarp** or **tetrasporagium**)?

5. What product forms as a result of meiosis in a red alga?

6. Which pigment is responsible for the red color in species of Rhodophyta?

7. Why do dried red algae lose their red color when rehydrated?

8. Examine the herbarium specimen of *Porphyra*. Is the thallus **haploid** or **diploid**? Carpospores will be generated on this thallus. Where will the carposporophyte occur? Explain. If the carpospores fall onto the proper substrate such as oyster shells to germinate into the **conchelis-phase** will the phase be **haploid** or **diploid**? Where will meiosis specifically occur in the life cycle of *Porphyra*?

9. Examine the formalin preserved and herbarium collections of *Corallina*. Notice that this alga is not fleshy as are other red alga. How does *Corallina* differ from other fleshy red alga and what is primarily responsible for this difference?

 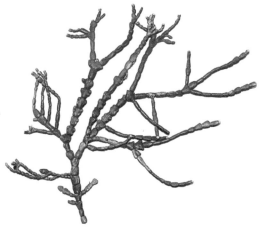

10. What type of life cycle do the Floridiophyceae express? Draw and diagram the typical life cycle of a Floridiophyceaen alga such as *Polysiphonia*. Identify **haploid** and **diploid** stages in the life cycle and include where **syngamy, meiosis**, and **mitosis** occurs.

11. Examine any and all demonstration material, including the "edible" algae! Differentiate between those which are red algae (Rhodophyta) and those which are brown algae (Phaeophyceae).

Readings

Hommersand, M.H. & Fredericq, S. 1990. Sexual reproduction and cystocarp development. *In:* Cole, K.M. & Sheath, R.G. (eds.), pp. 305-345, Biology of the Red Algae. Cambridge University Press, New York.

Kingsbury, J.M. 1969. Seaweeds of Cape Cod and the Islands. The Chatham Press Inc. Chatham, MA, USA.

Pueschel, C.M. 1989. An expanded survey of the ultrastructure of red algal pit plugs. Journal of Phycology 25:625-636.

Saunders, G.W. & Hommersand, M.H. 2004. Assessing red algal supraordinal diversity and taxonomy in the context of contemporary systematic data. American Journal of Botany 91:1494-1507.

Domain Eukarya
> **Supergroup Archaeplastida**
>> **Phylum Chlorophyta (Green Algae)**
>>> **Class Micromonadophyceae**
>>> **Class Chlorophyceae**

Order Volvocales	*Chlamydomonas, Pandorina, Volvox, Eudorina*
Order Chlorococcales	*Chlorella, Pediastrum, Hydrodictyon, Scenedesmus*
Order Chaetophorales	*Fritschiella*

>>> **Class Ulvophyceae**

Order Ulvales	*Ulva, Enteromorpha*
Order Cladophorales	*Cladophora*
Order Dasycladales	*Acetabularia*
Order Caulerpales	*Caulerpa, Halimeda, Codium*

>>> **Class Charophyceae**
>>>> **Order Zygnematales (Conjugatophyceae)**

Family Zygnemataceae	*Zygnema, Spirogyra*
Family Desmidiaceae	*Cosmarium, Closterium, Staurastrum, Micrasterias*
Order Coleochaetales	*Coleochaete*
Order Charales	*Chara, Nitella*

Related Terms: chlorophylls *a* and *b*, carotenoids, starch, 2-4+ whiplash flagella, cellulose, pectin, sometimes calcified, mostly fresh water, cellular coenocytic, siphonous (multicellular), asexual or sexual reproduction, volvocine series, coenobial, oogamous sexual reproduction, zygote, asexual daughter colony, non-flagellate unicells and colonies, heterotrichous, rhizoid, sporothallus, meiospores, gametothalli, isogametes, quadriflagellate zygote, giant diploid cells with multiple roles (zygote, sporothallus, meiosporangium, gametothallus, gametangium), cytoplasmic streaming, rhizoids, utricles, conjugation, conjugation tube, bilateral symmetry, zygote retention in gametothallus, jacket cells, precursors of land plants

Green Algae

The Chlorophyta primarily occur either in freshwater or marine habitats. Some do occur in terrestrial habitats on moist surfaces and in wet soil. The group is represented by more than 6,000 species. Species are unicellular, multicellular colonial, or multicellular filaments. Some of the multicelluar species such as sea lettuce (*Ulva lactuca*) form elaborate membranous sheets while others such as mermaid's wineglass (*Acetabularia* sp.) form intricately dissected calcified thalli. The Chlorophyta also include many species that are extremely important to science as they have served as model organisms for developmental, physiological and evolutionary research (*Acetabularia, Chara, Chlamydomonas,* and *Volvox*) that has greatly enriched our understanding of biology today. Both molecular and morphological data suggest

that the green algae, in part, gave rise to the land plants. Imagine our Earth without the land plants. The primary constituent of cell walls is cellulose. Like the red algae some groups are calcified. Chlorophyll *a* and *b* which give members of this group their distinctive green coloration are derived from a primary endosymbiosis of a cyanobacteria. The food reserve is mostly amylose and amylopectin starches.

1. Examine slide 78 labeled *Chlamydomonas* which is a flagellated unicellular green alga. Label the **flagella, cell wall, nucleus, pyrenoid, chloroplast, starch,** and **contractile vacuole**.

2. Examine slide 79 labeled *Volvox* w.m. *Volvox* is, at specific points in its life cycle, a macroscopic "large" colonial organism. Look for examples of both **vegetative and sexual reproduction**. How many flagella are associated with each cell?

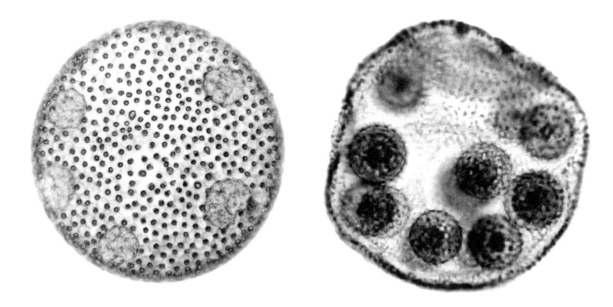

3. Examine the living culture of *Spirogyra*. Identify the **cell wall**, **nucleus**, **chloroplast**, **pyrenoid**, and **vacuole**. Where are the chloroplasts located in the cell?

4. Examine slide 80 labeled *Spirogyra* scalariform conjugation w.m. (scalariform = ladder-like). Members of this class lack motile gametes. Instead, fertilization results from a process known as conjugation. Conjugation involves pairing of compatible cells, formation of conjugation tubes between them, and fusion of the two cytoplasms to form a zygote. Identify **conjugating cells**, the **conjugation tubes**, **zygotes**, and the **ploidy level** of the various structures you see in this slide. What type of life cycle does this organism exhibit? Which filament is **male** which filament is **female**?

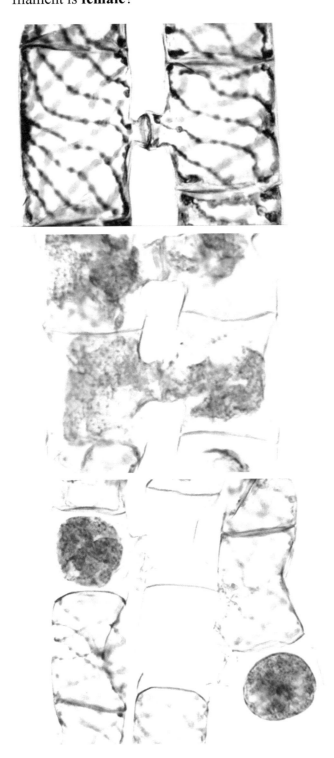

5. Examine slide 81 labeled *Zygnema* vegetative and conjugation w.m. Identify the two **stellate chloroplasts** each with a **central pyrenoid**. The **nucleus** occurs within a **cytoplasmic bridge** between the two stellate chloroplasts. Compare/contrast the basic morphology of *Zygnema* with *Spirogyra*. Where does conjugation occur in *Zygnema*? Identify **haploid** and **diploid** structures.

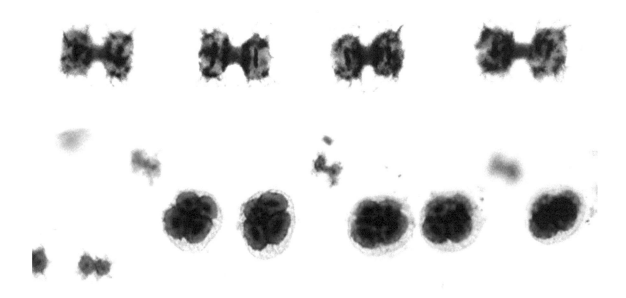

6. Examine slide 83 labeled *Ulva* thallus. *Ulva* (sea-lettuce) has a parenchymatous thallus, (evidence that cell division is proceeding in more than one plane). The thallus is, in reality, two cell layers thick (identify). *Ulva* is a common green alga on seashores in the temperate zone and is of interest because it has isomorphic alternation of generations, that is, that the haploid gametophytes are indistinguishable from the diploid sporophytes.

153

7. Examine slide 84 labeled *Enteromorpha*. *Enteromorpha* has a crinkled tubular thallus. Like *Ulva* it is common on seashores in the temperate zone. It too has isomorphic alternation of generations. Some phycologists consider the two genera to be synonymous. Here we will consider the two to be separate genera. Is the thallus you are viewing **haploid** or **diploid**? Explain.

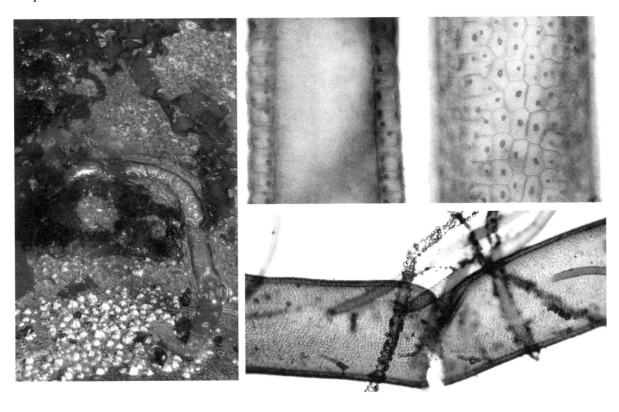

8. Draw and diagram the life cycle of *Ulva* below. Identify **haploid** and **diploid stages** in the life cycle and include where **syngamy, meiosis**, and **mitosis** occurs.

9. Examine slide 82 labeled Desmids mixed. The Desmids are nonflagellate, have unicellular thalli, often with an intricate configuration and are bilaterally symmetrical. Desmids are common freshwater planktonic algae that are bright green in color and easily distinguished from other freshwater planktonic algae. Note the distinctive bilobed nature of each cell and any variations in general cell shape. Check the slide for *Cosmarium, Closterium, Microasterias, and Staurastrum.* Do you see any other algal type on this slide?

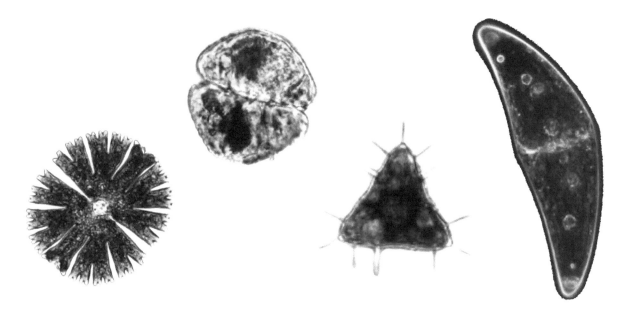

10. Examine slide 85 labeled *Pandorina* sphere or the live culture. How many cells make up the sphere? What term is collectively used to refer to the sphere? Are these cells **haploid** or **diploid**? Are these cells **flagellated** or **not**?

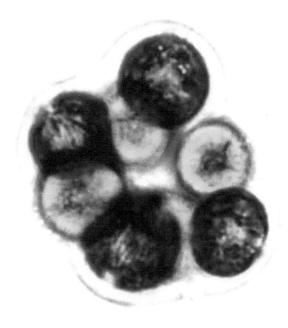

11. Examine slide 86 labeled *Eudorina* hollow sphere or the live culture. How many cells make up the sphere? What term is collectively used to refer to the sphere? Are these cells **haploid** or **diploid**? Are these cells **flagellated** or **not**?

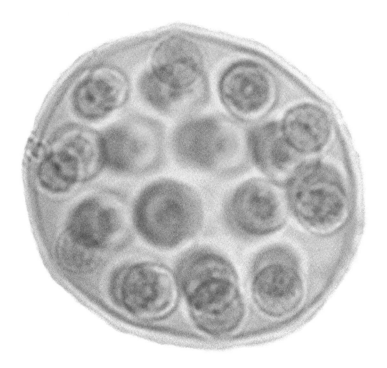

12. Examine slide 87 labeled *Hydrodictyon* or the living culture. How does the habit of *Hydrodictyon* compare with that of *Eudorina* and *Pandorina*.

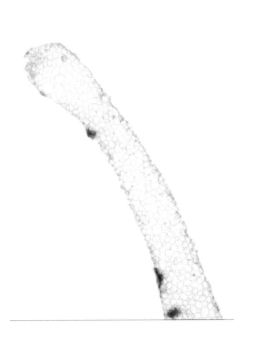
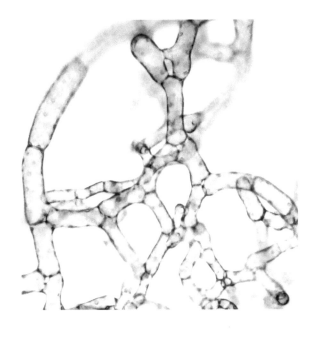

13. Examine slide 88 labeled *Nitella* section and slide 89 labeled *Nitella* sex organs or the living culture. Identify the **oogonium**, the **antheridium**, and **spirally arranged protective cells**. Is the vegetative thallus **haploid** or **diploid**? Where does meiosis occur?

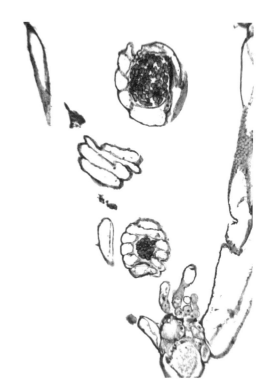

14. Examine slide 90 and 91 labeled *Chara* sex or the living culture. Identify the **oogonium**, the **antheridium**, and **spirally arranged protective cells**. Is the vegetative thallus **haploid** or **diploid**? Where does meiosis occur?

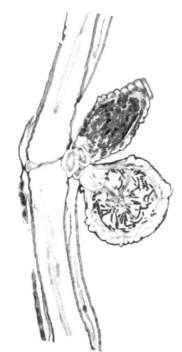

157

15. Draw and diagram the life cycle of *Chara* below. Identify **haploid** and **diploid stages** in the life cycle and include where **syngamy, meiosis**, and **mitosis** occurs.

16. Examine slide 92 labeled *Coleochaete* w.m. Identify an **encysted zygote** if present. Note and explain the difference in the habit of *Coleochaete* as compared to *Chara* and *Nitella*.

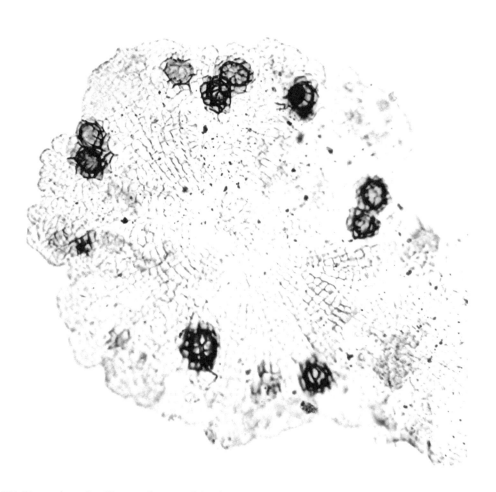

17. Examine the live culture of *Pediastrum* and *Scenedesmus*. How many cells make up the plate? What term is collectively used to refer to the plate? Are these cells **haploid** or **diploid**? Are these cells **flagellated** or **not**?

Readings

Bold, H.C. & Wynne, M.J. 1985. Introduction to the Algae: Structure and Reproduction. 2nd ed. Prentice-Hall, Inc., Englewood Cliffs, NJ, USA.

Kingsbury, J.M. 1969. Seaweeds of Cape Cod and the Islands. The Chatham Press Inc. Chatham, MA, USA.

Lewis, L.A. & McCourt, R.M. 2004. Green algae and the origin of land plants. American Journal of Botany 91: 1535-1556.

Norris, J.N. 2010. Marine algae of the northern gulf of california: Chlorophyta and Phaeophyceae. Smithsonian Institution Scholarly Press, Washington, D.C., USA.

Smith, G. M. 1950. The fresh-water algae of the. United States. 2nd ed. McGraw-Hill Book Company, Inc., New York, NY, USA.

Taylor, W.R. 1957. Marine Algae of the Northeastern Coast of North America. 2nd ed. University of Michigan Press, Ann Arbor, MI, USA.

Tilden, J. 1910. Minnesota Algae V1: The Myxophyceae of North America and adjacent regions. University of Minnesota Press. Minneapolis, Minnesota, USA.

Whitford, L.A. & Schumacher, G.J. 1969. A manual of the fresh-water algae in North Carolina. The North Carolina Agricultural Experimental Station, NC, USA.

Domain Eukarya
    Supergroup Archaeplastida
        Phylum Marchantiophyta (Liverworts)
            Class Marchantiopsida (Complex-thalloid)
            Subclass Marchantiidae
                Order Marchantiales   *Conocephallum, Marchantia, Riccia*
            Class Jungermanniopsida (Leafy & Simple-thalloid)
            Subclass Pelliidae (Simple-thalloid)
            Subclass Metzgeriidae (Simple-thalloid)
            Subclass Jungermanniidae (Leafy)
                Order Jungermanniales   *Plagiochila*
        Phylum Anthocerotophyta (Hornworts)   *Anthoceros*
        Phylum Bryophyta (Mosses)
            Class Bryopsida (True mosses)   *Polytrichum, Mnium*
            Class Sphagnopsida (Peat mosses)   *Sphagnum*
            Class Andreaeopsida (Granite Mosses aka Lantern Mosses)

Related Terms: gametophyte, sporophyte, protonema, venter, archegonium, antheridium, neck, zygote, gemmae, gemmae cup, rhizoids, columella, archegoniophore, antheridiophore, seta, capsule, paraphyses, operculum, calyptra, biflagellate sperm, peristomal teeth, foot suspensor, hydroids

Bryophytes

Bryophytes are the non-vascular land plants. Historically this common name was applied to the Bryophyta which at that time included the liverworts, hornworts and mosses. Today these groups are placed in their own phyla, that is, the Marchantiophyta, the Anthocerotophyta, and the Bryophyta, respectively. To varying degrees these all grow in moist habitats and are anatomically simple for land plants. Both molecular and morphological data suggests that these were the first groups to diverge from main stream land plants. The liverworts however represent the most ancestral lineage. Each group is monophyletic. Though similar in that each lack advanced conducting structures and lignins, differences among them are striking (Table 1). Approximately 19,000 species of bryophytes have been described. Of these, mosses represent half of the diversity. The hornworts are the least diverse group and are represented by about 100 species. Unlike all of the other land plants, the bryophytes have a dominant gametophyte. The sporophyte is not only ephemeral, but completely dependent upon the gametophyte for its development and survival.

Table 1. Summary of representative bryophyte characteristics.

	liverwort	hornwort	moss
Characteristic			
Rhizoid			
color	hyaline (clear)	hyaline (clear)	brown
septate	absent	absent	present
Chloroplast	many per cell	single	many per cell
Oil bodies	present	absent	absent
Gametophyte stomata	air pore (in few)	absent	absent
Leaf			
morphology	divided, undivided, or absent	absent	undivided
arrangement	rows or absent	absent	spiral
Capsule			
columella	absent	present	present
elater/pseudoelater	present	present	absent
operculum	absent	absent	present
stomates	absent	present	present
spore perine	absent	absent	present
Mutualism (with *Nostoc*)	present (in a few)	present (in some)	absent

Phylum Marchantiophyta

Liverworts are the most primitive of the land plants and are represented by 6,000-8,000 species. These are divided into the simple thalloid liverworts (Metzgeriidae), leafy liverworts (Jungermanniidae), and complex thalloid liverworts (Marchantiidae). The liverworts are somewhat unique among the bryophytes in that they may reproduce asexually through the production of gemmae. Some of the dioicous (see below) complex thalloid liverworts produce antheridiophores and archegoniophores. Based on your etymological abilities what would you predict to be produced on each of these?

Terms:

dioicous - a term applied by bryologists when male and female gametangia occur on separate gametophytes.

dioecious - a term applied by botanists when male and female gametangia occur on separate sporophytes in vascular plants and separate gametophytes in the algae.

monoicous - a term applied by bryologists when male and female gametangia occur on the same gametophyte.

monoecious - a term applied by botanists when male and female gametangia occur on the same sporophyte in vascular plants and the same gametophyte in the algae.

gemmae - an asexual bud-like propagule as occurs in the liverworts capable of developing into a new individual

carpocephalum - the sporogonial receptacle in some liverworts like those in *Marchantia*

archegoniophore - an upright structure consisting of a stalk and cap upon which archegonia are borne

antheridiophore - an upright structure consisting of a stalk and cap upon which antheridia are borne

163

1. Examine slide 101 labeled *Marchantia* life cycle. Identify the **sporophyte**. Identify the **gametophyte**. What term is best applied to the red encapsulated structures in the image?

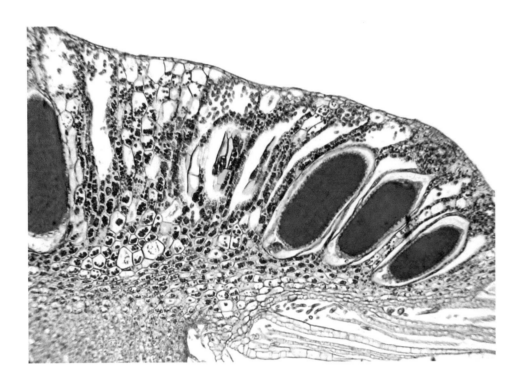

2. Examine slide 186 labeled *Marchantia* archegonia l.s. Identify the **archegoniophore**, **archegonia, carpocephalum, eggs, venter,** and **neck cells.** Which of these are **haploid** and which are **diploid**?

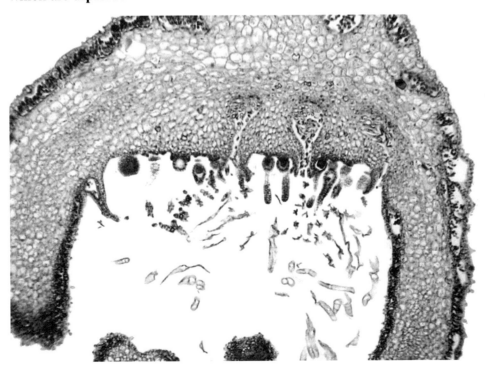

164

3. Examine slide 184 labeled *Marchantia* sporophyte l.s. Identify the **capsule, seta, spores elaters,** and the **foot** of the **sporophyte**. Identify the sporophyte. Identify the **gametophyte**. Where is the sporophyte located on this complex thalloid liverwort?

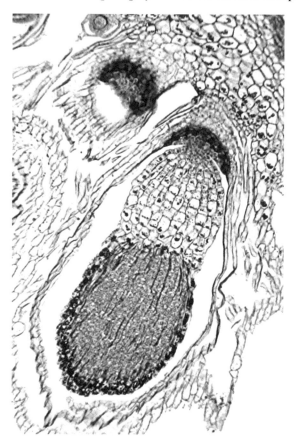

4. Examine slide 187 labeled *Marchantia* cupule l.s. Identify the **sporophyte**. Identify the **gametophyte**. What term is best applied to the reddish-brown encapsulated structures? Are these **haploid** or **diploid**? Where are these structures located?

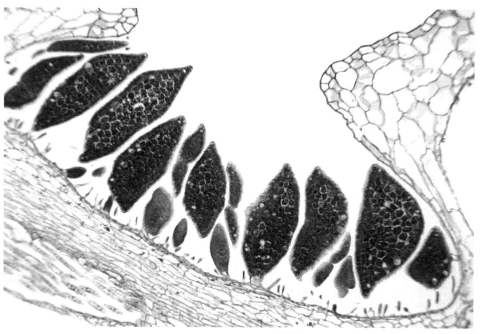

5. Draw and label the life cycle of a complex thalloid liverwort such as *Marchantia* below. Identify **haploid** and **diploid stages** in the life cycle and include where **syngamy, meiosis,** and **mitosis** occurs.

6. Prepare and examine a wet mount of the aquatic liverwort *Riccia fluitans*.

7. Examine slide 102 labeled *Riccia natans* mature sporophyte l.s. (this species is currently known as *Ricciocarpus natans*). Identify the **sporophyte**. Identify the gametophyte. Label the **capsule, seta, spores** and **elaters.**

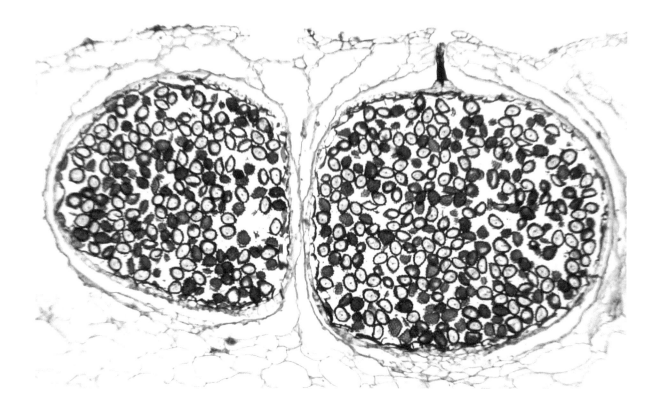

8. Prepare and examine a wet mount of the terrestrial liverwort *Porella*.

Phylum Anthocerotophyta

Hornworts are the most recently derived group of bryophytes and are represented by less than 15 genera and approximately 100 species. Although anatomically similar to some of the simple thalloid liverworts, they do differ from these and the mosses in that their sporophyte is horn-like in appearance. Additional similarities and differences among these groups are summarized in Table 1.

9. Examine the living hornwort specimen.

Phylum Bryophyta

The mosses are represented by approximately 12,000 species that are divided among the true mosses, peat mosses, granitic mosses, and *Takakia*. True mosses (Class Bryopsida) and peat mosses (Class Sphagnopsida) are common throughout North America. Granitic mosses (Class Andreaeopsida), which include only two genera, occupy rocky habitats from tropical to arctic climates. *Takakia*, which includes only two species, occurs in western North America and central and eastern Asia. Like the gametophores of liverworts the gametophytes of many mosses have limited conductive tissue differentiation. Additional similarities and differences among the bryophytes per se are summarized in Table 1.

10. Examine slide 103 labeled *Mnium* life cycle. Identify the **protonema**. Is the protonema **haploid** or **diploid**? From which structure does the protonema arise? What does the protonema give rise to?

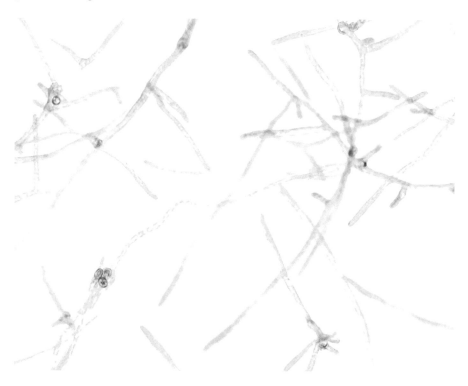

11. Examine slide 189 labeled *Mnium* leaf x.s. Identify the **hydroids**. Is the leaf **haploid** or **diploid**?

12. Examine slide 192 labeled *Mnium* archegonial head l.s. Identify the **archegonia** and **egg**. Did the egg arise through **meiosis** or **mitosis**?

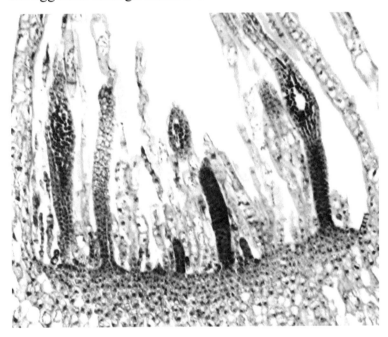

13. Examine slide 191 labeled moss antheridial head l.s. Identify the **antheridia** and **sperm**. Are the antheridia part of the **gametophyte** or **sporophyte**? Did the sperm arise through **meiosis** or **mitosis**?

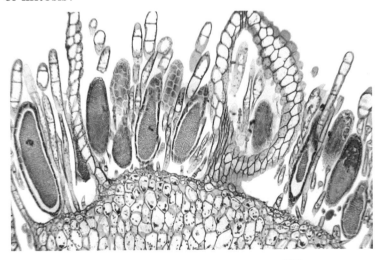

14. Examine the gametophyte and sporophyte of a true moss in the Order Plasticales.

15. Examine the living material of *Polytrichum*. Identify the **gametophyte**, **sporophyte**, **capsule**, and **calyptra**. Is the calyptra **haploid** or **diploid**? Does the calyptra have **fine hairs** associated with it?

16. Examine the living material of a true moss for sporophytes. Identify the **capsule, operculum (cap), calyptra, stalk** and **peristomal teeth**. Are peristomal teeth **haploid** or **diploid**? What is the function of peristomal teeth?

17. Examine slide 156 labeled *Polytrichum* mature capsule median l.s. Identify the **sporophyte**. Identify the **gametophyte**. Label the **spores, columella, capsule,** and **operculum (cap)**.

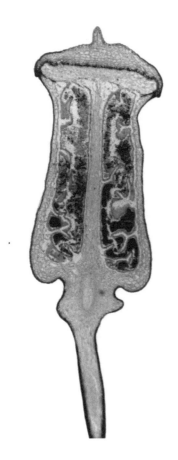

18. Examine slide 157 labeled *Polytrichum* capsule and calyptra c.s. Identify the **sporophyte, gametophyte**, **spores, columella, capsule,** and **operculum (cap)**.

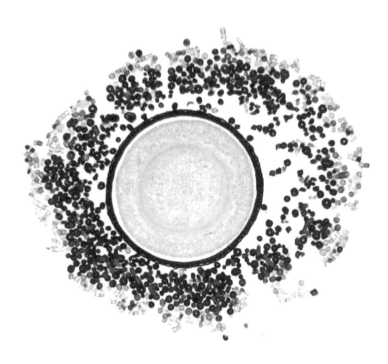

19. Examine the living specimen of *Atrichum*. Make a wet mount of a leaf. Identify, **marginal cells**, the **serrate margin**, **median leaf cells**, and **photosynthetic lamella**.

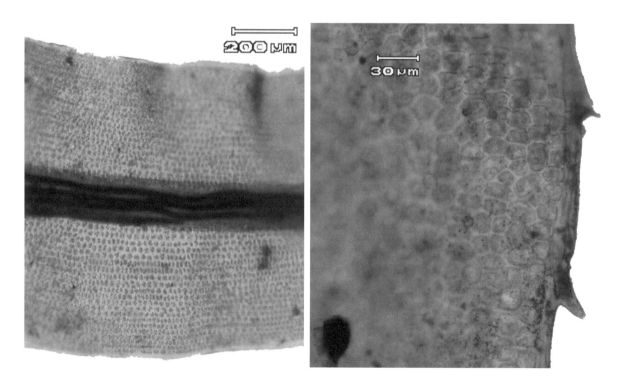

20. Draw and label the life cycle of the moss *Polytrichum* below. Identify **haploid** and **diploid** stages in the life cycle and include where **syngamy, meiosis**, and **mitosis** occurs.

21. Examine the living culture of *Sphagnum*. Prepare a wet mount of a leaf. You can also examine slide 193 labeled *Sphagnum* leaf w.m. Note the photosynthetic cells (chlorophyllose cells) as well as the dead cells (hyaline cells). What is the function of the **hyaline cells**? Is this a **gametophyte** or **sporophyte**?

22. Examine slide 195 labeled *Sphagnum* antheridia l.s. as well as the living culture. Identify the **sperm cells, antheridial jacket cells, antheridia,** and the **leaves** of the **gametophyte**.

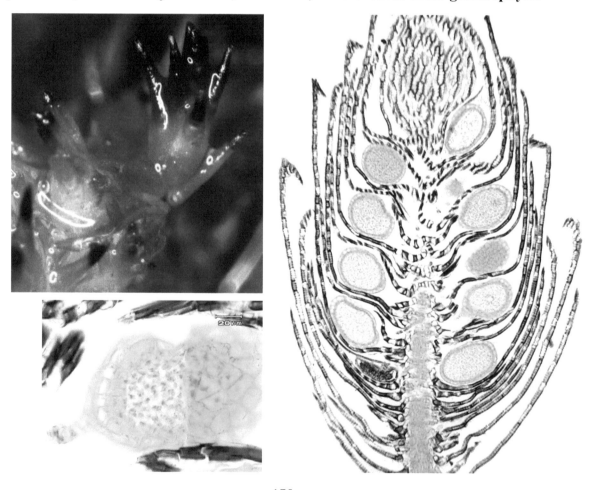

23. Examine slide 194 labeled *Sphagnum* mature sporophyte as well as the live culture. Identify the **pseudopodium, foot, capsule wall, spores,** and **operculum**. Is the pseudopodium **haploid** or **diploid**?

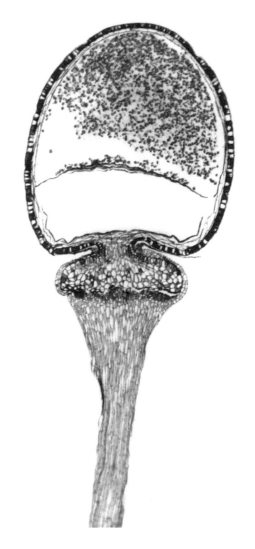

Readings

Anderson, L.E., Shaw, A.J., & Shaw, B. 2009. Peatmosses (Sphagnum) of the southeastern United States. New York Botanical Garden Press, New York.

Crandall-Stotler, B., Stotler, R.E., & Long, D.E. 2009. Phylogeny and classification of the Marchantiophyta. Edinburgh Journal of Botany 66: 155-198.

Goffinet, B. & Shaw, A.J. 2009. Bryophyte Biology, 2nd ed. Cambridge University Press, Cambrige.

Hébant, C. 1977. The conducting tissue of bryophytes. J. Cramer, Vaduz.

Shaw, A.J. & Renzaglia, K.S. 2004. Phylogeny and diversification of bryophytes. American Journal of Botany 91: 1557-1581.

Domain Eukarya
    Supergroup Archaeplastida
        Phylum Lycopodiophyta (Club Mosses)
            Order Protolepidodendrales   *Asteroxylon*
            Order Lepidodendrales   *Lepidodendron, Sigillaria*
            Order Lycopodiales
                Family Lycopodiaceae   *Lycopodium*
            Order Selaginellales
                Family Selaginellaceae   *Selaginella*
            Order Isoetales
                Family Isoetaceae   *Isoetes*

Related Terms: micropyll, sporophyll, strobilus, strobilophore, homosporous, tetrad of spores, bisexual gametophyte, endophytic symbiotic fungi, rhizoids, ligule, heterospory, exospory, endospory, microsporangia, microsporophylls, megaspores, microspores, megasporangia, megasporophylls, trabeculae, velum, archegonia, antheridia, zygote, foot, suspensor, biflagellate sperm, sterile sporophyll, hollow root, rhizophore, cambial layer, 4 megaspores versus extreme heterospory, parichnos scars

Club Mosses

Phylum Lycopodiophyta

Although the club mosses are most abundant in the tropics and wet temperate regions of the world they are a cosmopolitan group that is represented by nearly 1,200 species, the most common of which occur in the genera *Lycopodium* and *Selaginella.* Both molecular and morphological data suggest that the Lycopodiophyta are the first lineage of extant (i.e. living) vascular plants to diverge from main stream tracheophytes (i.e.vascular plants) and hence are an ancestral group of the vascular plants. The Lycopodiophyta as a group arose in the Devonian Period (410-360 Ma) and greatly

diversified during the Carboniferous Period (360-240 Ma). Some of the club moss species were very similar to those we see today whereas others were arboreal (tree-like), reaching 30 m in height and forming extensive forests during the Carboniferous Period. Sexual reproduction is oogamous where water acts as a conduit for gamete transfer. The dominant and conspicuous stage in the life cycle (heteromorphic alternation of generations) is the sporophyte which contains vascular tissue in the form of xylem. The sporophyte possesses microphylls (leaves with a single unbranched vascular trace) that may be photosynthetic and vegetative or reproductive. If reproductive the microphyll is more appropriately called a sporophyll. The sporophylls may be grouped together in the apex of a branch to form a structure called a strobilus. Sporangia are derived from either a sporophyll cell (foliar) or a branch cell (cauline). Depending upon the specific group the sporangia may produce either one or two types of spores (homospory and heterospory, respectively). In heterosporous plants megasporangia will give rise to large female spores (megaspores) and the microsporangia generate the smaller male spores (microspores). Spores of *Lycopodium* are used in the making of a substance called *Lycopodium* powder. Historically it was used as a flash powder in photography and also in making fireworks. Today it is used theatrically as a "safe flash" in magic acts.

Table 1. Summary of representative club moss characteristics.

	club moss	spike moss	quillwort
Characteristic			
Leaf type	isophyllous	isophyllous or anisophyllous	isophyllous
Spore condition	homosporous	heterosporous	extreme heterospory
Sporophyll ligule	absent	present	present
Gametophyte development	exosporal	endosporal	endosporal
Secondary growth	absent	absent	present

1. Examine the herbarium or live specimens of local *Lycopodium* species. Identify the **stem, rhizome**, **roots**, **microphylls,** and **sporophylls**. What is the difference between a microphyll and a sporophyll?

2. Sporangia may occur singly as in *Lycopodium lucidum* or may be grouped in a strobilus. The strobilus/strobili may be sessile as in *Lycopodium obscurum* or occur on a long stalk as in *L. complanatum*. Examine the herbarium and live specimens of representative *Lycopodium* species. Identify and label the **sporophylls, microphylls, sporangia, strobilus,** and **strobilophore.**

3. Examine the living material of a strobiloid (cone-like reproductive structure) *Lycopodium.* Identify the **sporophylls, microphylls,** and **strobilus**.

4. Examine slide 115 labeled *Lycopodium lucidulum* stem c.s. Identify the **type of protostele**. Do you see **sclerenchyma**? Identify the **microphylls** and **leaf traces**.

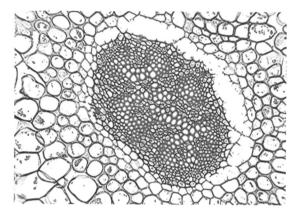

5. Examine slide 196 labeled *Lycopodium* rhizome c.s. Identify the **type of protostele**. Do you see **sclerenchyma**? Explain. Identify the **microphylls** and **leaf traces**. Compare the anatomy of this section with the stem section above.

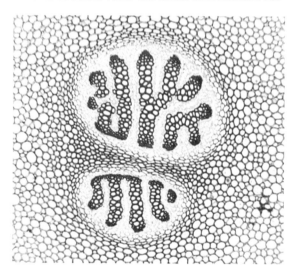

6. Examine slide 116 labeled *Lycopodium lucidulum* root c.s. Identify the **type of protostele**. Do you see **sclerenchyma**? Explain. Identify the **microphylls** and **leaf traces**. Compare the anatomy of this section with the stem and rhizome sections above.

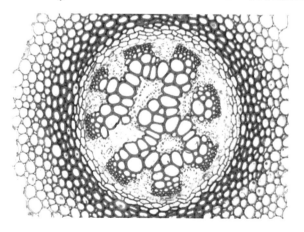

7. Examine slide 117 labeled *Lycopodium lucidulum* strobilus c.s. Identify the **sporophylls, sporangia, microphylls,** and **spores**.

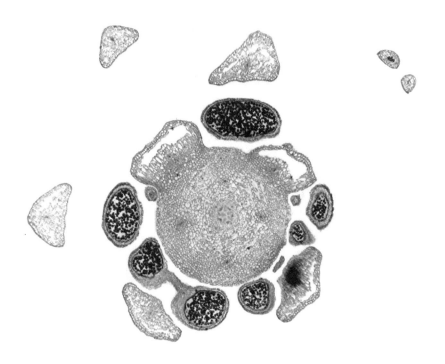

8. Examine slide 118 labeled *Lycopodium lucidulum* strobilus l.s. Identify the **sporophylls, sporangia, microphylls, strobilus,** and **spores**.

9. Examine slide 119 labeled *Lycopodium lucidulum* gemma c.s. Some species of *Lycopodium* form vegetative structures termed gemmae, or bulbils, which become detached from the plant and grow into new sporophytes. These structures arise in the positions of microphylls and consist of an organized bud and preformed root primordia. The factors that favor bulbil formation are not well understood. Identify the **bulbils**.

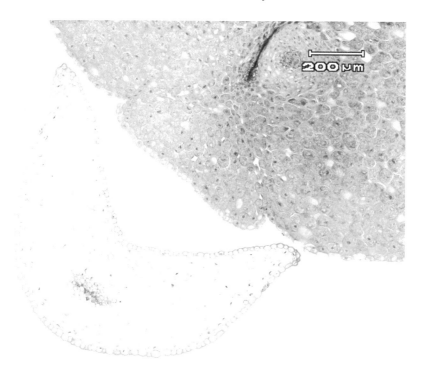

10. Examine the living material of *Selaginella* and identify **stems** and **microphylls** below. Is the specimen shown below **isophyllus** or **anisophyllus**? Explain.

11. Examine slide 121 labeled *Selaginella* stem c.s. Identify the **meristele**, **central air-filled canal**, **trabeculae**, **microphylls**, and **leaf traces**. The trabeculae are formed by elongate endodermal cells.

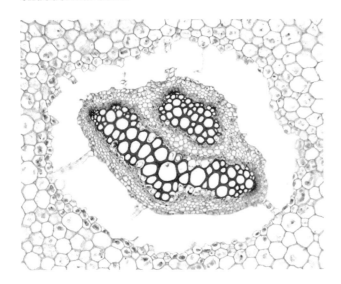

12. Examine slide 120 labeled *Selaginella* strobilus l.s. Identify the **microphylls**, **sporophylls**, **megasporophylls**, **megasporangia**, **megaspores**, **microsporophylls**, **microsporangia**, **microspores**, and **ligules**. Is *Selaginella* **homosporous** or **heterosporous**?

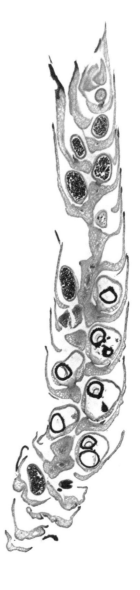

13. Examine slide 122 labeled *Selaginella* sporeling. The young sporophyte is attached to the gametophyte, showing root, stem, and first pairs of microphylls. Identify the **gametophyte,** and the **root, stem,** and **microphylls** of the sporophyte.

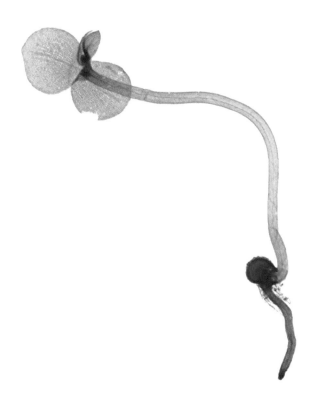

14. Examine slide 123 labeled *Isoetes* stem l.s. and slide 124 labeled *Isoetes* stem c.s. Identify the **sporophylls, sporangia, spores, corm-like rhizome, trabecula, ground tissue,** and **vascular tissue.**

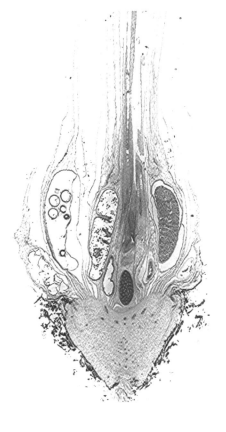

15. Examine slide 125 labeled *Isoetes* leaf c.s. Note the large **air chambers**.

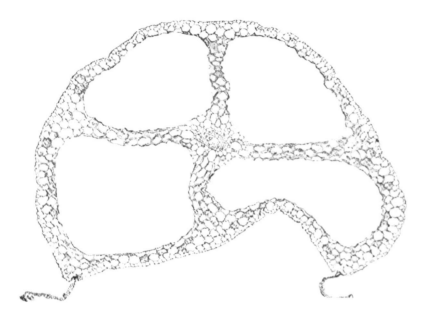

16. Examine slide 126 labeled *Isoetes* root c.s. The root consists of a cylindrical cortex which surrounds a large air cavity and a vascular cylinder which is supported in a flange of the cortex in the cavity. The primary xylem and phloem are collateral (opposite each other) in arrangement, the phloem being oriented toward the cavity. The air cavity is formed by a breakdown of cortical cells throughout the length of that portion which has emerged from the rhizophore.

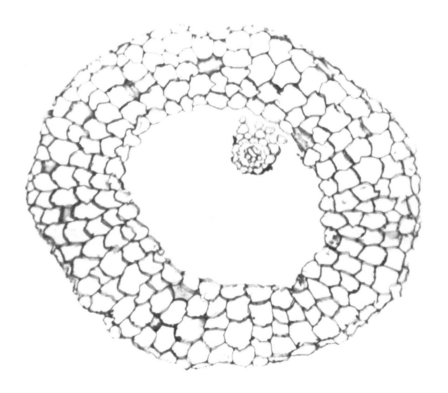

17. Examine slide 128 labeled *Isoetes* microsporophylls and megasporophylls l.s. Identify the **microspores**, **microsporangia**, **velum**, **ligules**, **sterile sporophylls**, **microsporophylls**, **megasporophylls**, **megasporangia**, **megaspores**, **trabecula**, **roots** (probably better examples than slide 126), and **leaf traces**.

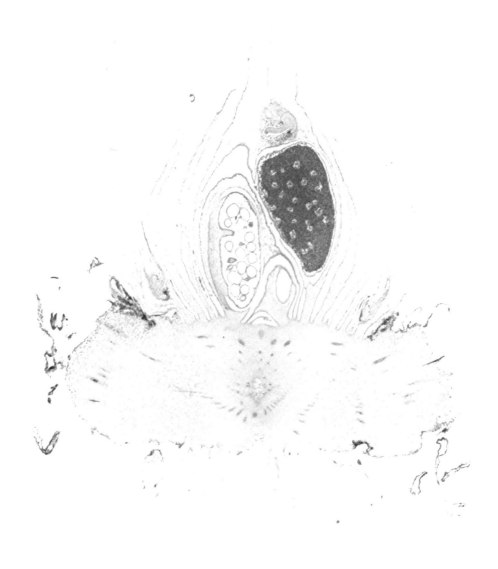

18. Examine the lycophyte fossils.

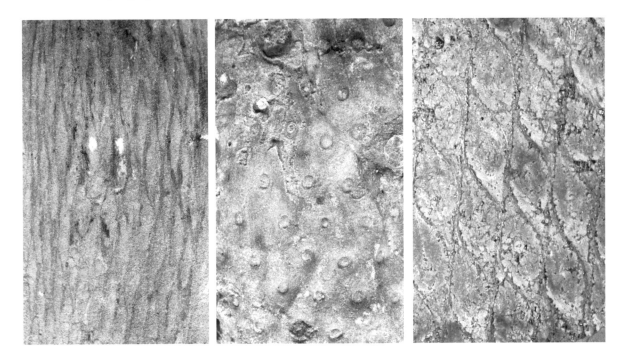

Readings

Andrews, H.R. 1961. Studies in Paleobotany. John Wiley and Sons, Inc., New York, USA.

Bocheński, T.A. 1939. On the structure of Sigillarian cones and the mode of their association with their stems. Polish Academy of Science 7:1-16.

Felix, C.J. 1952. A study of arborescent lycopods of southeastern Kansas. Annals of the Missouri Botanical Garden 39: 263-288.

Maden, A.R., Renzaglia, K.S., & Whittier, D.P. 1996. Ultrastructure of the spermatozoid of *Lycopodiella obscurum* (Lycopodiaceae). American Journal of Botany 83: 419-429.

Øllgaard, V. 1987 A Revised Classification of the Lycopodiaceae *sensu lato*. Opera Botanica 92: 153-178.

Renzaglia, K.S., Bernhard, D.L., & Garbary, D.J. 1999 Developmental ultrastructure of the male gamete of *Selaginella*. International Journal of Plant Science. 160: 14-28.

Whittier, D.P. & Webster,T.R. 1986. Gametophytes of *Lycopodium lucidulum* from axenic culture. American Fern Journal 76: 48-55.

Wikström, N. & Kenrick, P. 1997. Phylogeny of Lycopodiaceae(Lycopsida) and the relationships of *Phylloglossum drummondii* Kunze based on *rbc*L sequences. International Journal of Plant Science. 158: 862-871.

Domain Eukarya
 Supergroup Archaeplastida
  Phylum Monilophyta (Ferns and Horsetails)
   Class Psilotopsida
    Order Psilotales (Whisk Fern)   *Psilotum*
    Order Ophioglossales (Grape Fern)   *Botrichium,*
                     Ophioglossum

Related Terms: dichotomous branching, terminal sporangia, sterile branches, lateral sporangia, overtopping, protostele, sporangia, synangia, synangium, eusporangiate

   Class Equisetopsida (Horsetails)
    Order Equisetales   *Equisetum*
    Order Sphenophyllales   *Sphenophyllum*
    Order Calamitales   *Calamites,*
                    Annularia

Related Terms: ribs of sclerenchyma, pith canal, vallecular canals, carinal canals, strobilus, sporangiophore, sporangium, appendicular tip, homosporous, archegonium, antherdium, gametophyte, sporophyte, peltate sporangiophore, fused microphylls, jointed stem, unique lateral branch growth, organ genus

   Class Polypodiopsida (Leptosporangiate Ferns)
    Order Filicales
     Family Osmundaceae
    Order Marsileales
    Order Salviniales

Related Terms: macrophylls, megaphylls, leptosporangiate, homosporous, heterosporous, exospory, endospory, non-circinnate vernation, circinnate vernation, bisexual gametophyte, fertile pinnule, synangium, fertile/sterile pinnae, rachilla, sorus, annulus, sori, false indusium, reniform indusium, peltate indusium, sporocarp, megasporocarp, microsporocarp, sporophore

   Class Marattiopsida (Marattoid Ferns)
    Order Marattiales

Related Terms: megaphylly, eusporangiate, homospory, endospory, circinnate vernation, bisexual gametophyte, synangial sorus, compound fronds

Ferns and Horsetails

Phylum Monilophyta (Pteridophyta)

The ferns are a diverse assemblage of land plants that includes the true ferns (Class Polypodiopsida), whisk and snake-tongue ferns (Class Psilotopsida), king ferns (Class Marattiopsida), and horsetails (Class Equisetopsida). That said lets be very clear on one point, that is, that despite the fact that numerous classification schemes

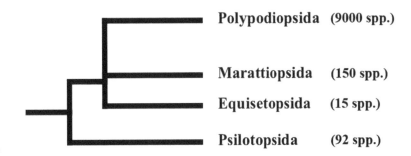

Polypodiopsida	**(9000 spp.)**
Marattiopsida	**(150 spp.)**
Equisetopsida	**(15 spp.)**
Psilotopsida	**(92 spp.)**

Figure 1. Proposed Phylogeny of the Monilophyta

have been proposed for this group there is little consensus among them. Some botanists still refer to the club mosses, horsetails, and whisk ferns collectively as the fern allies which of course is not a monophyletic group, but still a convenient way to address these organisms on the whole. All ferns have vascular tissue in the form of xylem and phloem, the dominant phase in the life cycle is the sporophyte generation, all produce spores, and all have flagellate sperm. The true ferns (including the king ferns and snake-tongue ferns) have maintained their level of species diversity since the Carboniferous Period (360-240 Ma). The whisk ferns were, and continue to be, an insignificant component of the flora. The horsetails, which were a codominant group in the Carboniferous Period with the lycopods, are an insignificant component of the flora today.

Class Psilotopsida

The Class Psilotopsida includes the Orders Psilotales (whisk ferns) and Ophioglossales (snake-tongue ferns). Whisk ferns consist of two genera (*Psilotum* and *Tmesipteris*) and approximately 10 species. The genus *Psilotum* occurs in tropical and subtropical woodlands worldwide as well as throughout the southern United States. The sporophyte is dichotomously branched and superficially resembles a leafless stem. It can generate an extensive subterranean rhizome from which new stems could arise. The stems possess scaly leaf-like structures (enations referred to as prophylls) that lack vascular tissue, unlike the megaphylls of the true ferns and microphylls of the club mosses and horsetails. However, *Tmesipteris*, the other genus within the Psilotales, does possess microphylls. *Psilotum* lacks true roots, but instead has a rhizome covered in rhizoids. Gametophytes are small,

190

subterranean, and more or less colorless. Sporophyte stems bear three fused sporangia (synangia) that form on the ends of short branches (therefore cauline in origin).

The Ophioglossales are represented by two families the Ophioglossaceae (snake-tongue ferns) and Botrychiaceae (moonworts). Species within the group are somewhat atypical 'ferns' in that they produce a single vegetative frond per year or none at all. Gametophytes are small, subterranean, and more or less colorless and may take up to two decades to form a sporophyte. The genus *Ophioglossum*, specifically *Ophioglossum reticulatum*, has the highest chromosome number of any plant species at $2n = 1260$.

The Psilotopsida arose in the middle of the Devonian Period (410-360 Ma) where, as a group, it was most diverse, but never a significant component of the flora. Today it still represents an inconspicuous component of our flora.

1. Examine the living specimen of *Psilotum*. Identify the **stem, prophylls, rhizoids, rhizome, synangia,** and **dichotomous branching**.

2. Examine slide 129 labeled *Psilotum* rhizome. Identify the **epidermis**, **cortex**, **phloem**, **xylem**, and **rhizoids**. Describe the arrangement of xylem and phloem. The stele is a simple kind of protostele. What term best describes this type of stele?

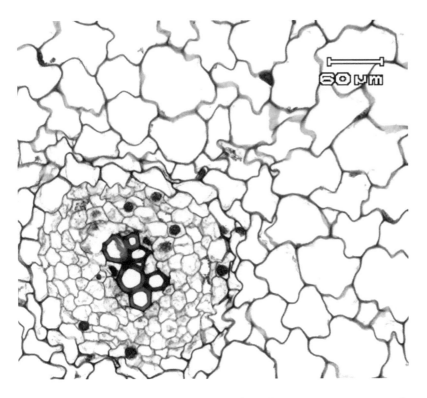

3. Examine slides 130 and 131 labeled *Psilotum* young sporangium c.s. and *Psilotum* sporangium l.s. respectively. Identify the **synangial wall** and **sporangia wall tissue**. How many sporangia are contained within the synangium?

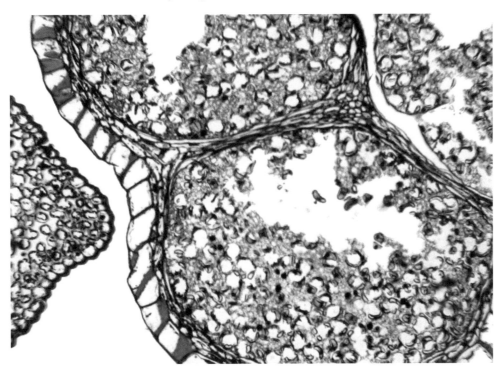

4. Examine slide 197 labeled *Ophioglossum* root, rhizome, stipe, c.s. Fern roots arise from the rhizome or upright stem near or below the leaf bases. They are usually small. Their epidermal cells are usually thin-walled. A generous portion of the inner cortex may be sclerenchymatous. Fern roots are monarch, diarch or tetrarch with few tracheids in the xylem. Describe the actinostele in your root.

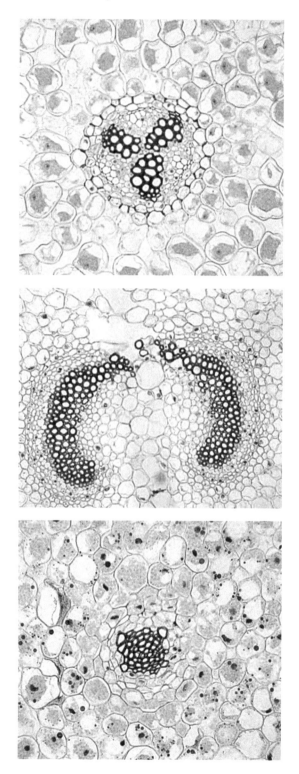

Class Equisetopsida (horsetails, scouring rushes)

The Equisetopsida, commonly referred to as the horsetails or scouring rushes, is represented by a single extant genus (*Equisetum*) with nearly 25 species. The sporophyte stem is obviously segmented and ridged. Some bear lateral branches while others do not. Reproductive stems in some species are achlorophyllous. Superficially the genus appears to lack leaves, but upon closer inspection, microphylls (fused), similar to those of the Lycophyta (club mosses) are evident. Epidermal cells contain silica that is in part responsible for their rigidness. Anatomically the stem has a series of cavities that extend throughout the long axis of the plant body. The central most of these is the central canal. Those associated with the conductive tissue and adjacent to the ridges are the carinal canals, whereas those between the ridges or in the valleys are the vallecular canals. Sporangia form on the terminal portion of a specialized branch called a sporangiophore and like the Psilotales are cauline in origin. Generally, when sporangia form, many sporangiophores bearing sporangia are generated. These sporangiophores form at the apex of the stem and produce a cone-like structure called a strobilus. Extensions of the spore cell wall (elater-bearers) aid in their dispersal.

As with the Lycophyta, extinct members of the Equisetopsida were large and arboreal (some nearly 30 m tall) occurring in low-land swamps during the Carboniferous Period (360-240 Ma). These extinct arboreal species in concert with the Lycophyta gave rise to our coal deposits. The Equisetopsida as a group arose in the middle of the Devonian Period (410-360 Ma) and reached the height of their diversity in the Carboniferous Period. Today they represent an inconspicuous component of the flora.

5. Examine herbarium specimens of *Equisetum arvense*. Identify the **leaf sheath**, **nodes**, **internodes**, **lateral branches**, **peltate shield**, **sporangia**, **segmentation**, **ridges**, and **strobili**.

6. Examine slide 141 labeled *Equisetum arvense* sterile stem c.s. Identify the **stomata, carinal canals, vallecular canals, pith cavity, central canal, cortex, epidermis, vascular bundles,** and **regions of sclerenchyma**.

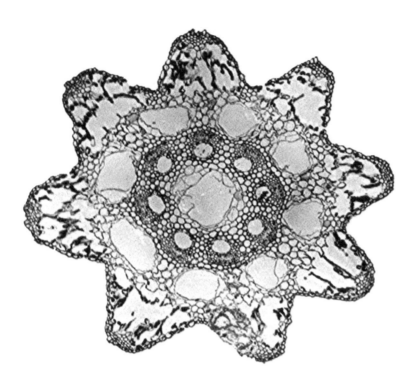

7. Examine slide 142 labeled *Equisetum arvense* rhizome c.s. How does the arrangement of cell types compare with that of the stem?

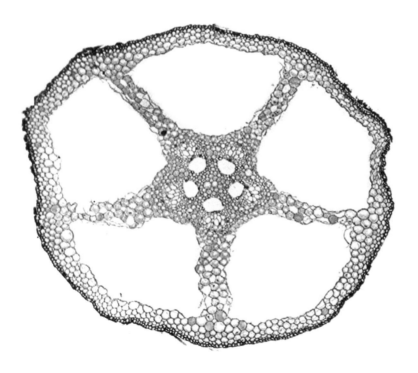

8. Examine slide 143 labeled *Equisetum intermedium* root c.s. The outer cortex is oftentimes sclerenchymatous whereas the inner cortex is oftentimes parenchymatous. Can you explain this anatomy? The xylem is triarch or tetrarch (diarch in smaller roots). Small roots have one large xylem element in the center of the root. Identify the **xylem**. Is this a **small** or **large** **root**?

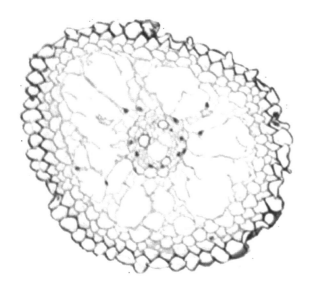

9. Examine slide 144 labeled *Equisetum arvense* branch apical cell. Identify the **branch apical cell(s)**.

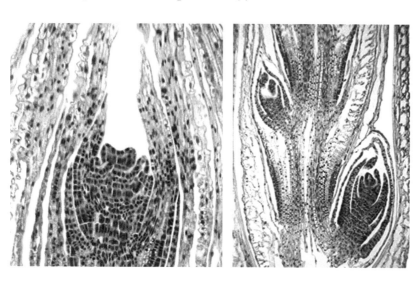

10. Examine slide 145 labeled *Equisetum arvense* mature strobilus median l.s. and c.s. Identify the **central axis**, **peltate shield**, **sporangia**, **sporangiophores**, and **homosporous spores**.

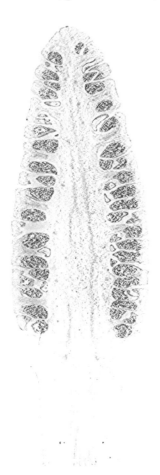
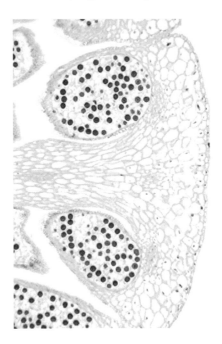

11. Examine slide 149 labeled *Equisetum arvense* spores and elaters w.m. Identify the **elater-bearers**. Why are these elater-bearers really not elaters or even pseudo-elaters as seen in the Bryophytes? Explain.

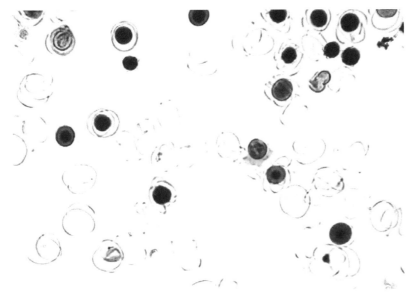

12. Examine slide 150 labeled *Equisetum* young prothallia. What is a prothallus? Identify **rhizoids** and the **prothallus**. Is the prothallus **haploid** or **diploid**?

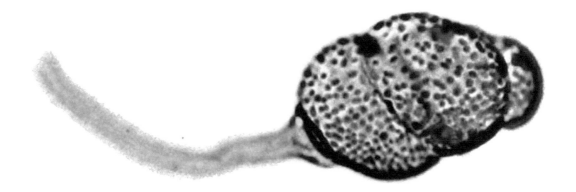

13. Examine slide 147 labeled *Equisetum arvense* gametophyte w.m. This slide is a more developed stage of the previous slide. Identify **rhizoids** and the **prothallus**. What shape does the prothallus have?

14. Examine slide 146 labeled *Equisetum* gametophyte antheridia w.m. Identify the **antheridia** with sperm, **archegonia,** and **rhizoids**.

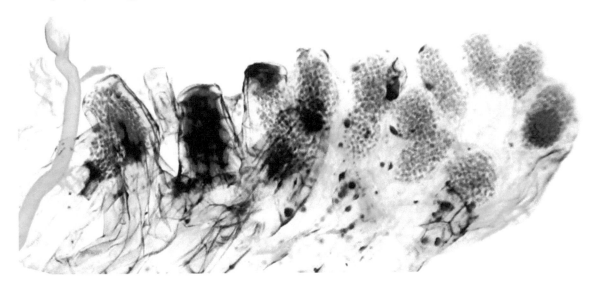

15. Examine slide 148 labeled *Equisetum* gametophyte w.m. developing sporophyte. Note that the sporophyte is attached to the gametophyte. Identify the **sporophyte, gametophyte,** and the **archegonia.**

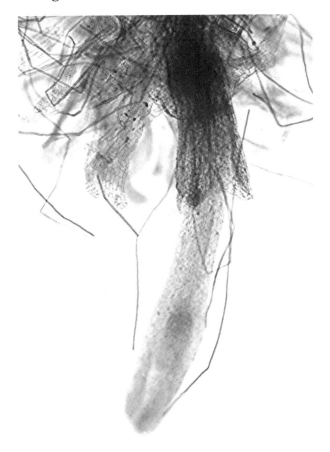

Class Polypodiopsida (True ferns)

The Polypodiopsida or true ferns include nearly 12,000 species some of which are arboreal (i.e. tree-like). Some ferns are evergreen while others are ephemeral. Gametophytes are heart shaped, bear clear rhizoids on the abaxial surface, and are inconspicuous. The most common phase of the life cycle is the sporophyte generation. The sporophyte consists of a rhizome (an underground stem), roots, and a megaphyll (frond). A

megaphyll by definition is a leaf with a branched vascular trace associated with a "break" or "gap" in the conducting tissues of the stem. The frond consists of a stipe (basal stalk) and a lamina (photosynthetic blade). The further extension of the stipe into the lamina is the rachis. The lamina can be simple or compound. Fronds are usually referred to as being compound or simple based on the dissection or lack thereof of the lamina. Compound lamina are once divided into units called pinnae or twice divided into units called pinnules which bear pinnae. The central axis of a pinnae is a costa. The folding and rolling of a frond in the bud stage is referred to as vernation. Some ferns, demonstrate erect vernation (folding along the long axis), whereas most demonstrate circinate vernation (rolling along the long axis). Spores may be alike (homosporous) or dimorphic (heterosporous) and are produced within sporangia that may occur on the underside of pinnae or solitary on a rachilla (a costa lacking pinnae). Groupings of sporangia are called sori. These may be naked or partially covered by a false or true indusium. Thick walled sporangia that develop from several epidermal cells are eusporangiate. These sporangia are sessile, large (> 0.5 mm in dia.), lack specialized cells for predetermined dehiscence, and produce large numbers of spores. Thin walled sporangia that develop from a single epidermal cell are leptosporangiate. These sporangia are stalked, small (< 0.1 mm in dia. – with some exceptions), bear specialized cells for predetermined dehiscence (annulus), and produce relatively few spores. The leptosporangiate ferns are more derived. The Polypodiopsida arose in the middle of the Devonian Period (410-360 Ma) and still represent a diverse component of the flora today.

16. Examine the demonstration material of living ferns and herbarium specimens. Identify the **frond, rachis, stipe, indusium, rhizome, roots, pinna, scales,** and **sori** of *Athyrium filix-femina*.

17. Examine the demonstration material of living ferns and herbarium specimens. Identify the **fertile frond, sterile frond, stipe, pinna, rachis,** and **sporangia** of *Onoclea sensibilis.* Note that fronds are either vegetative or reproductive.

18. Examine slide 162 labeled *Dennstaedtia* rhizome c.s. *Dennstaedtia* has a siphonostele which is an evolved form of a protostele. In a siphonostele the pith occurs in the center of the vascular cylinder and in this case consists of centrally located fibers. Phloem and endodermis occur on both sides of the xylem. Pith parenchyma occurs between the innermost endodermis and the pith fibers. Indentify the **pith fibers, pith parenchyma, endodermis, pericycle, phloem, xylem, cortex,** and **epidermis.**

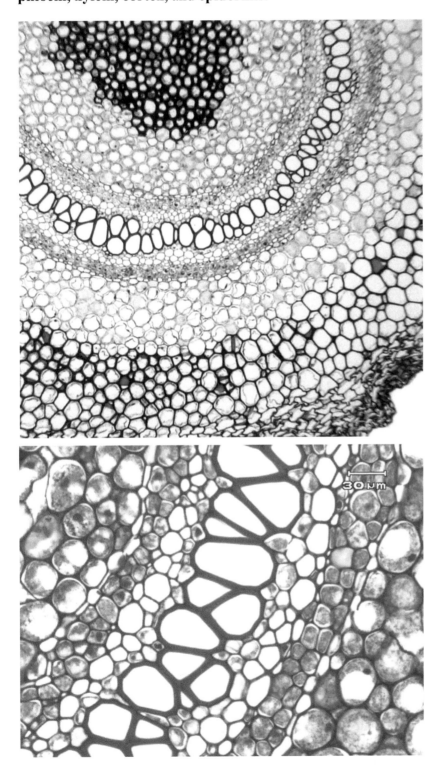

19. Examine slide 161 labeled *Polypodium* rhizome c.s.   A siphonostele in which many overlapping leaf gaps occur at the same level is specifically a dictyostele or a dissected siphonostele. Each discrete bundle of vasculature is a meristele. Each meristele is surrounded by an endodermis and pericycle. This is the most highly evolved stele in the Polypodiopsida. How many meristeles make up your dictyostele?   Indentify the **pith parenchyma, endodermis, pericycle, phloem, xylem, cortex, epidermis,** and individual **meristeles.**

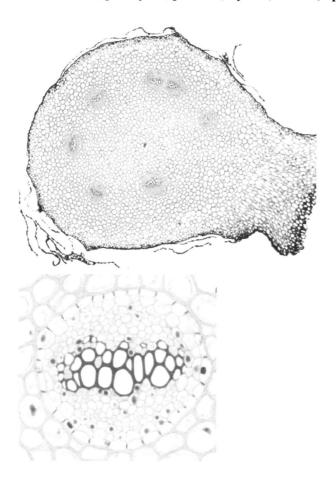

20. Examine slide 198 labeled *Polypodium* leaflet. Identify the **sporangia**, **spores**, **annulus**, and **sorus**. Do you see either an **indusium** or **false indusium**?

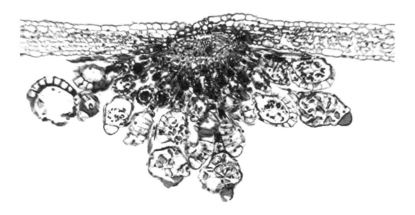

21. Examine slide 199 labeled *Cyrtomium* sporangia w.m. Identify the **annulus**.

22. Examine slide 212 labeled *Cyrtomium falcatum* sorus on leaf l.s. Identify **sporangia**, **spores**, and the **indusium.**

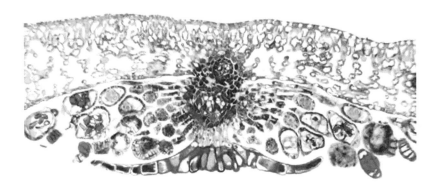

23. Examine slide 104 labeled fern prothallium antheridia l.s. Identify the **antheridia** and **sperm**. Is the prothallus **haploid** or **diploid**?

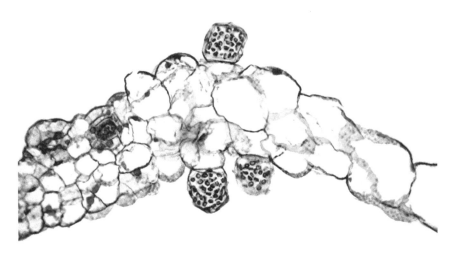

24. Examine slide 105 labeled fern archegonia l.s. Identify the **archegonia** and **egg**. Which cellular process gives rise to the antheridia: **mitosis** and **cytokinesis**, **meiosis** and **cytokinesis** or **syngamy**?

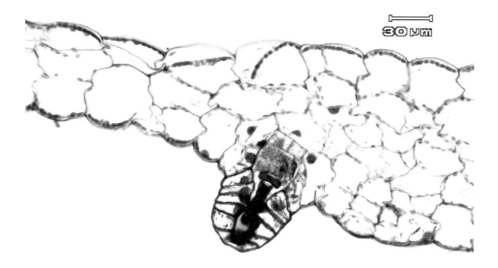

25. Examine slide 107 labeled fern prothallium fertilization. Identify the **antheridia, sperm,** and **rhizoids**. Identify **haploid** and **diploid** tissue.

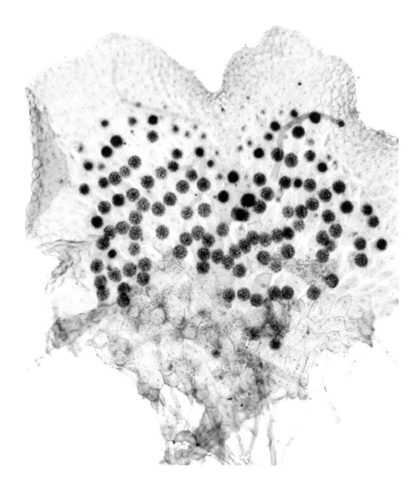

26. Examine slide 108 labeled fern prothallium embryo. Identify the **embryo**, **antheridia,** and **archegonia**?

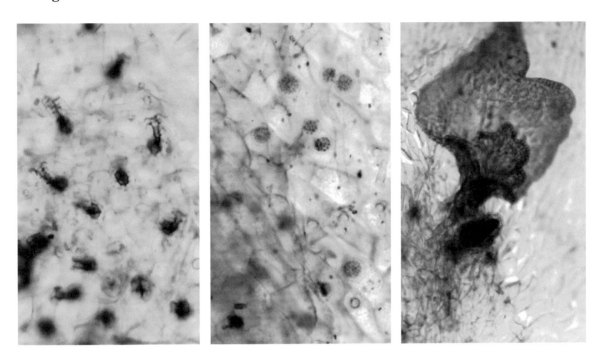

27. Examine slide 106 labeled fern prothallium young sporophyte. Identify **haploid** and **diploid** tissue.

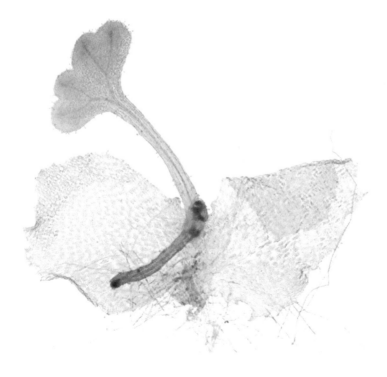

28. Examine the living demonstration material of the aquatic fern *Marsilea*. Identify the **compound leaf, sporocarp,** and **sporocarpophore**.

29. Examine slide 112 labeled *Marsilea* leaflet c.s. The anatomy of *Marsilea* leaves is dependent upon whether the leaf is floating, submerged, or aerial. The mesophyll of submerged leaves is not differentiated into both spongy and palisade mesophyll. The stomata of floating leaves are sunken and only occur on the adaxial surface. Was this leaf **submerged, floating,** or **terrestrial**?

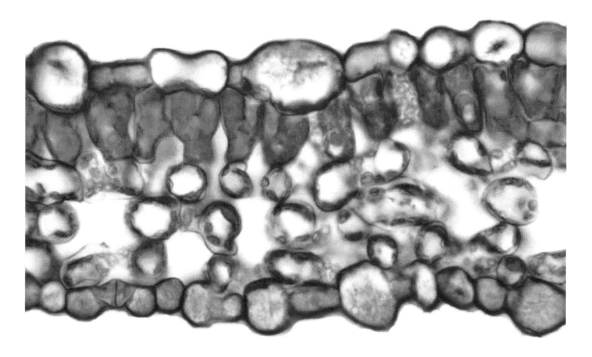

30. Examine slide 111 labeled *Marsilea* rhizome c.s. The prostrate rhizome of *Marsilea* is siphonostelic, specifically an amphiphloic siphonostele (ring of phloem, xylem, phloem). The inner portion of the cortex usually consists of compact tissue, the outer portion is lacunate with larger air spaces around radiating rows of parenchyma. The rhizomes of submerged plants generally have a thin-walled parenchymatous pith, whereas those growing on mud or damp soil have a sclerotic pith. Where did your plant grow?

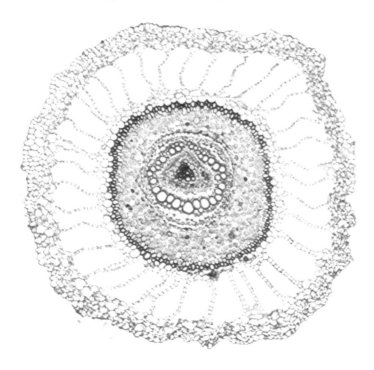

31. Examine slide 113 labeled *Marsilea* sporocarp c.s. Identify the **sporocarp wall**, **megaspore**, **megasporangium**, **microspore**, **microsporangium**, and **gelatinous sporophore**. Which end of the sporocarp was attached to the sporocarpophore?

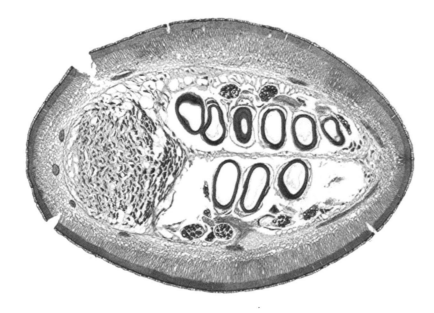

32. Examine living demonstration material of the aquatic fern *Salvinia*. Identify the **two green floating leaves, their hairs, creeping stem,** and **single finely dissected petiolate root-like leaf**. Describe the hairs.

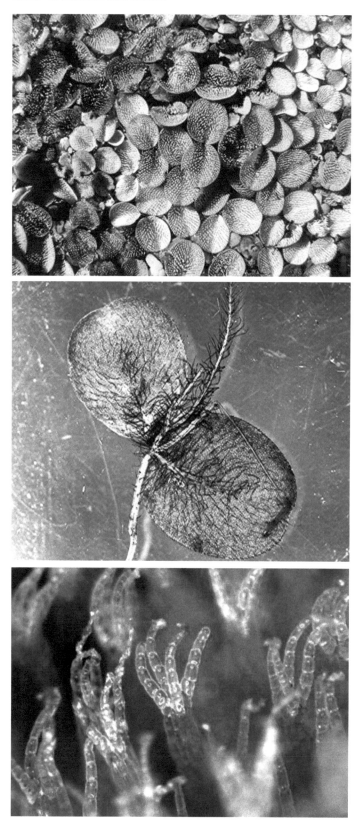

33. Examine demonstration material of the aquatic fern *Salvina molesta* for microsporocarps and megasporocarps. Although large quantities of elongated chains of sporocarps may be produced the sporangia within are generally devoid of spores because meiotic irregularities in this pentaplod species typically prevents the formation of spores.

34. Examine slide 109 labeled *Salvinia* sporocarp l.s. Identify **megasporocarps, megasporangia, megaspores, microsporocarps, microsporangia,** and **microspores**.

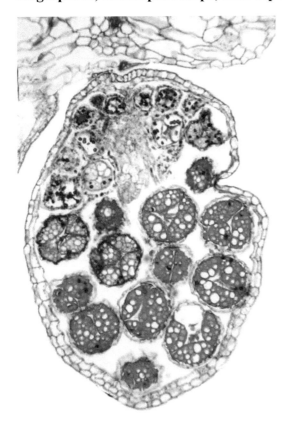

35. Examine living demonstration material of the aquatic fern *Azolla*. Describe the leaves and their arrangement.

36. Examine slide 110 labeled *Azolla caroliniana* sporocarps. Is this a **megasporocarp** or a **microsporocarp**?

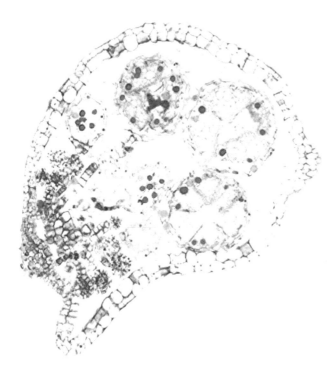

Class Marattiopsida

The marattiopsid or king ferns include approximately 60 extant species and diverged early from the mainstream fern line thereby forming a sister group with the equisetopsids. Gametophytes are large, liverwort-like and commonly mycorrhizal. Usually only one zygote develops per gametophyte. The fronds emerge from the dorsal surface of the gametophytes in contrast to their ventral emergence in the polypodiopsid ferns. The fronds of the marattiopsid ferns exhibit circinate vernation and are large (6 to 9 meters in length and 4.5 m wide) and complex (sometimes up to four times pinnate). The sporangia of these eusporangiate ferns are typically fused thereby forming synangia. The stipes of the fronds are thick, fleshy and erect. Recounting the story of the HMS Bounty and the bread fruit, Captain Bligh introduced the marattiopsid fern *Angiopteris erecta* from Tahiti to Jamaica where it too was served as a staple food for slaves.

37. Examine slide 213 labeled *Marattia* rhizome c.s. What type of stelar arrangement occurs in the marattiopsid ferns?

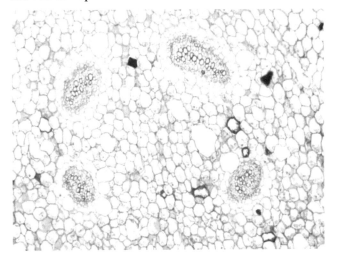

38. Examine slide 214 labeled *Marattia* leaflet/sporangia attached c.s. and slide 215 labeled *Marattia* mature sporangia. Identify the differences.

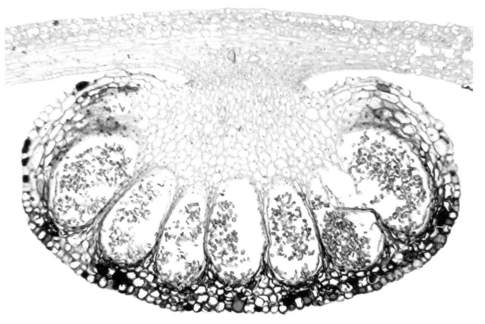

39. Examine any and all demonstration material including the fern and horsetail fossils.

Readings

Duckett, J.G. & Duckett, A.R. 1980. Reproductive biology and population dynamics of wild gametophytes of *Equisetum*. Botanical Journal of the Linnean Society 79: 205-210.

Holloway, J.E. 1939. The gametophyte, embryo and young rhizome of *Psilotum triquetrum* Schwarz. Annals of Botany 3: 313-336.

Keating, R.C. 1968. Trends of specialization in the stipe anatomy of *Dennstaedtia* and related genera. American Fern Journal 58: 126-140.

Loyal, D.S. & Grewal, R.K. 1966. Cytological study on sterility in *Salvinia auriculata* Aublet with a bearing on its reproductive mechanism. *Cytologia* 31: 330-338.

Mitchell, D. S. 1979. The incidence and management of *Salvinia molesta* in Papua New Guinea. Office of Environmnet & Conservation & Department of Primary Industry, P.N.G. I-VI, 1-51, f. 1-2, 15 pl.

Mohlenbrock, R.H. 1999. The Illustrated. Flora of Illinois. Ferns, 2nd ed. Southern Illinois University Press, Carbondale, IL, USA.

Pfeiffer, W.M. 1907. Differentiation of sporocarps in *Azolla* – Contributions from the Hull Botanical Laboratory 105. Botanical Gazette 44: 445-454.

Pryer, K.M., Schuettpelz, E., Wolf, P.G., Schneider, H., Smith, A.R., & Cranfill, R. 2004. Phylogeny and evolution of ferns (monilophytes) with a focus on the early leptosporangiate divergences. American Journal of Botany 91: 1582-1598.

Schneider, E.L. & Carlquist, S. 1999. SEM studies on vessels in ferns. 11. *Ophioglossum*. Botanical Journal of the Linnean Society 129: 105-114.

Smith, A.R., Pryer, K.M., Schuettpelz, E., Korall, P., Schneider, H., & Wolf, P.G. 2006. A classification for extant ferns. Taxon 55: 705-731.

Stiles, W. 1910. The structure of the areal shoots of *Psilotum flaccidum*. Annals of Botany 24: 373-387.

Webb, E. 1975. Stem anatomy and phyllotaxis in *Ophioglossum petiolatum*. American Fern Journal 65: 87-94.

Domain Eukarya
    Supergroup Archaeplastida
        Phylum Cycadophyta (Cycads)
            Class Cycadopsida
                Order Cycadales
                    Family Cycadaceae
                    Family Stangeriaceae
                    Family Zamiaceae

Related Terms: Spermatophyte, dioecious, manoxylic (wood high in parenchyma) not pycnoxylic (wood rarely has parenchyma), stems with girdling leaf traces, pollen sac, pollen cone scale, prothallial, antheridial, generative, tube, sterile stalk, fertile body, multiflagellate sperm, microsporangium, micropyle, pollination, pollen droplet, pollen chamber, ovule integument, nucellus, megasporocyte, microsporocyte, archegonial chamber, chalazal end, free-nuclear stage, proembryo

        Phylum Ginkgophyta (Maidenhair Trees)
            Order Ginkgoales
                Family Ginkgoaceae   *Ginkgo biloba*

Related Terms: Spermatophyte, dioecious, long shoots, spur shoots, dichotomously veined, catkin-like pollen cones, paired ovules on peduncle, fleshy seeds, Maidenhair tree, distribution restricted, cultivated, wind pollination, strong apical dominance, auxin, differential anatomy (long versus short shoot) pollen dispersal in four-celled state, female gametophyte development as in the cycads, fleshy, stony, papery layers, butyric acid, satellite chromosomes (female), salnuts

        Phylum Gnetophyta (Gnetophytes)
            Order Gnetales   *Gnetum*
            Order Ephedrales   *Ephedra*
            Order Welwitschiales *Welwitschia*

Related Terms: compound microstrobilus, compound megastrobilus, vessel elements, non-flagellated sperm, fleshy seeds, leaves dicot-like

        Phylum Pinophyta (Conifers)
            Order Cordaitales   *Cordaites*
            Order Lebachiales
            Order Pinales
                Family Taxaceae   *Taxus, Torreya, Austrotaxus, Paleotaxus*
                Family Podocarpaceae
                Family Cephalotaxaceae
                Family Cupressaceae *Juniperus, Thuja, Taxodium, Sequoia, Sequoiadendron, Metasequoia*
                Family Araucariaceae *Wollemia*
                Family Pinaceae   *Pinus, Abies, Larix, Picea Tsuga, Pseudotsuga*

Related Terms: compound ovuliferous cone, simple polleniferous cone, hard pines, soft pines, monoecious, pollination, 4-celled gametophyte, air bladder, megasporocyte, megagametophyte development, microgametophyte development, primary bract, ovuliferous scale, simple polyembryony, cleavage polyembryony, sterigmata, deciduous, quadrangular, petiolate, sessile, fleshy fused scales, dimorphic leaves, knees, General Sherman, General Grant, aril, taxol

Gymnosperms

Two major evolutionary trends that contributed to the dominance of the land plant lineage were the acquisition of pollen and seeds. Pollen (microgametophyte) allowed for sexual reproduction to occur in the absence of water. Recall that in the non-seed plants water was the conduit between male and female gametangia which facilitated sexual reproduction. Unlike spores, seeds provide nourishment to the embryo. This association increases the survival value of the young embryo and likely led in part to the dominance of the seed plants.

The seed plants (spermatophytes) are usually divided in two main groups, the non-flowering seed plants and flowering seed plants. The non-flowering seed plants are also known as the gymnosperms (from the Greek *gymnos* and *sperma* for naked seed). Gymnosperm seeds are born 'naked' and therefore not encapsulated. The group includes less than 1,000 extant species distributed among four phyla, the Cycadophyta (cycads), Ginkgophyta (ginkgos), Pinophyta (conifers), and Gnetophyta (gnetophytes). The flowering plants are also known as the angiosperms (from the Greek *angeion* and *sperma* for vessel and seed). Angiosperm seeds are born encapsulated. The group is represented by approximately 250,000 extant species. The Phylum Magnoliophyta has traditionally included two Classes, the Magnoliopsida (dicots) and Liliopsida (monocots). Today, the dicots are commonly divided into the paleodicots, magnoliids, and eudicots.

In seed plants, unlike the non-seed plants (bryophytes, lycophytes, & pteridiophytes), the propagule or dispersal unit of sexual reproduction is the seed, not the spore. The seed is an embryonic sporophyte inside a protective seed coat that is derived from parental sporophyte tissue. Usually nutritive tissue, the endosperm, is also present. In gymnosperms, the endosperm is derived from female gametophyte tissue only. In angiosperms, the endosperm is derived from a secondary fertilization event. In the seed plants the male gametophyte, or pollen, is transported by some means to the female gametophyte, or megagametophyte, which is contained within the ovule integument. The ovule integument will differentiate into the seed coat following fertilization. The gametophytes of seed plants are much smaller in size than are those of non-seed plants, even microscopic.

The Pollen Cone

Pollen is produced in pollen sacs (microsporangia) borne in varying numbers on the underside of pollen cone scales (microsporophylls), which vary in appearance depending upon the taxon. Pollination is facilitated by the wind and typically occurs in the spring or early summer. Vast amounts of pollen are produced.

The Immature Ovulate Cone

The female gametophyte develops from a single megaspore that lacks a spore wall. The megasporangium consists of a few cells contained within a layer of protective cells, the ovule integument. The whole complex of tissues is the ovule. Access to the gametophyte is via a gap in the integument called the micropyle. The ovules of *Zamia* (Cycadophyta) are borne in pairs on ovuliferous scales attached to the cone axis. The ovules of *Ginkgo* (Ginkgophyta) are truly naked and occur in pairs on the end of a dichotomous branched peduncle. The ovules of gnetophytes (Gnetophyta) occur in the axis of bracts or are  surrounded by a ring of tissue. The ovules of *Pinus* (Pinophyta) are borne in pairs on the upper surface of an ovuliferous scale that represents the only remaining structure of a once complex branch that arises from the primary cone axis above a cone scale. The cone scale or bract is typically not evident in mature cones of the Pinophyta, but can be seen in those of Douglas fir.

The Seed

Gymnosperm pollination, followed by fertilization may occur within the same year, or fertilization may occur in the subsequent year. If the latter occurs, pollen rests for nearly a year inside the pollen chamber having entered via the micropyle. During the second spring microgametogenesis resumes with pollen tube development and when it reaches the archegonium, fertilization may be facilitated.

Phylum Cycadophyta

The cycads include three living families Cycadiaceae, Stangeriaceae, and Zamiaceae, with 10 genera and approximately 250 species. Each bears pinnately compound leaves that are thick, leathery, and distinctly palm-like. Their trunks are stout. Today the cycads are a minor component of tropical and subtropical floras, but during the Jurassic Period (205-140 Ma) represented at least 20 percent of the land plant landscape. As such, the Jurassic Period is the 'Age of the Cycads'.

1. Examine slide 132 labeled *Zamia* stem c.s. Note the prominent leaf traces. Do you see the vascular cylinder and pith?

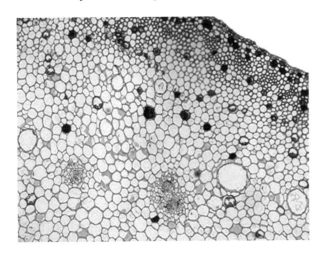

2. Examine slide 133 labeled *Zamia* root c.s. Note the formation of lateral roots. Where do these arise?

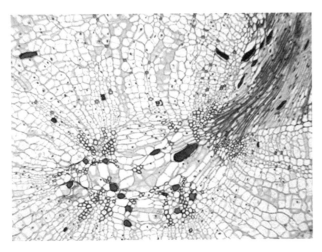

3. Examine slide 136 labeled *Zamia* leaflet c.s. Identify the row of **subepidermal fibers** and the **sunken stomates**.

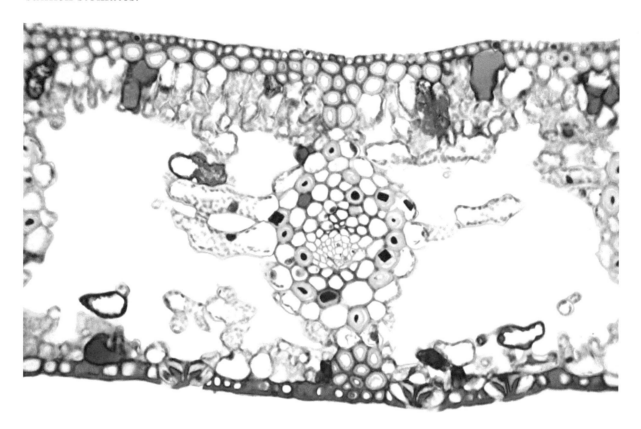

4. Examine slide 134 labeled *Zamia* male cone axis, c.s. Identify the **pollen cone scale, pollen sac (microsporangium),** and **pollen**.

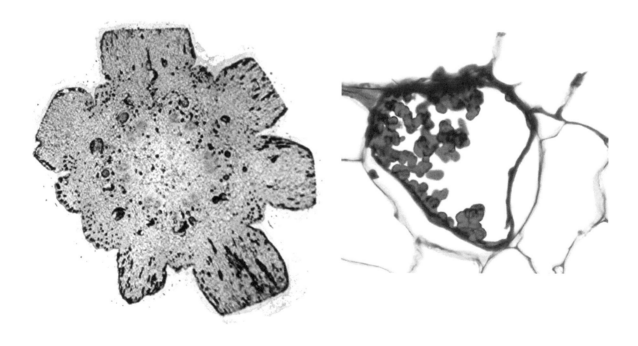

5. Examine slide 135 labeled *Zamia* archegonium l.s. Identify the vegetative tissue of the **female gametophyte (= endosperm after fertilization), micropyle, pollen chamber, nucellus (megasporangium), archegonial chamber, ovule integument, egg,** and **archegonium**.

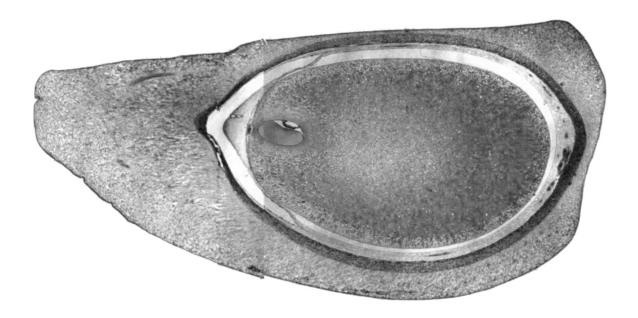

6. Examine slide 140 labeled *Zamia* mature embryo l.s. Identify the **endosperm, the cotyledons** (there two in most groups of the Cycadophyta), **the shoot,** and **root apices**.

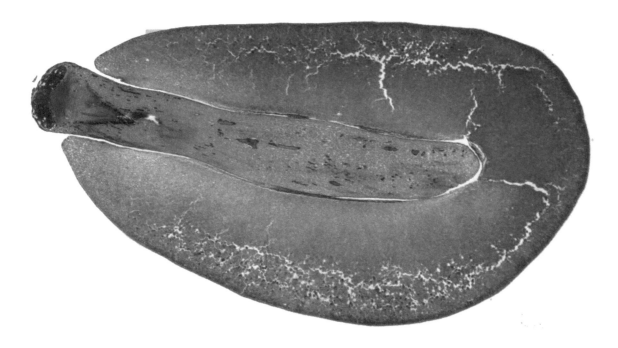

7. Examine slide 138 labeled *Cycas*: leaflet c.s. How does the anatomy of a *Cycas* leaflet compare to a *Zamia* leaflet? Identify **palisade mesophyll**.

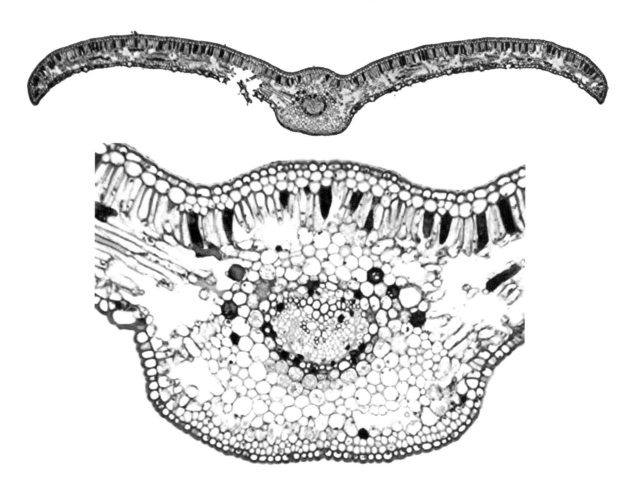

8. Examine slide 139 labeled *Cycas*: male sporophyll. Identify the **pollen cone scale, pollen sac (microsporangium),** and **pollen**.

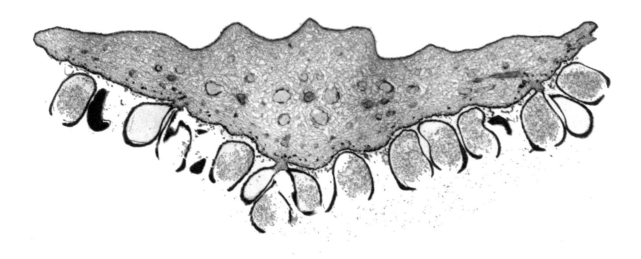

Phylum Ginkgophyta

Fossil evidence for the Ginkgophyta dates back to the Permian Period (290-245 Ma) and suggests that members of the phylum were widely distributed throughout the world, especially the northern hemisphere. Today only a single genus, *Ginkgo*, which dates back to the Jurassic Period (205-140 Ma), is represented by a single species, *G. biloba*, the maidenhair tree. The fact that this group has only one extant species makes it a monotypic member of the phylum. Western civilization was introduced to *G. biloba* in the 1690's by the German botanist Engelbert Kaempfer who was living in eastern China at the time. Eastern civilization was quite familiar with the species because it was cultivated on the grounds of most Buddhist temples and Shinto shrines at the time. So, *G. biloba* only exists today as a living taxon due to its cultivation in botanical gardens of temples

and shrines of China. The species has been historically planted in urban landscapes because of its air pollution tolerance. Their closest living relatives are the cycads. Based on gross morphology the two phyla are quite distinct. Like cycads *Ginkgo* is dioecious, but is distinguished by its thick, dichotomously veined leaves and apricot-like 'fruit' from which it gets its Chinese name *yin-hing* (silver apricot). The species has been touted by homeopaths to increase cognitive function, but scientific data does not support this claim.

9. Examine the demonstration material of *Ginkgo biloba*. Draw the leaves and note their bilobed nature and dichotomous venation. Where do the leaves arise?

10. Examine slide 151 labeled *Ginkgo* stem c.s. Identify the **dermal layers, cortex, pith, vascular tissue, resin ducts,** and **cell types**.

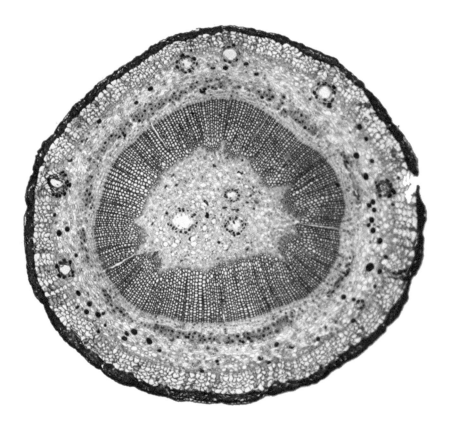

11. Examine slide 152 labeled *Ginkgo* root, c.s. Identify **dermal layer, cortex, vascular tissue,** and **cell types**.

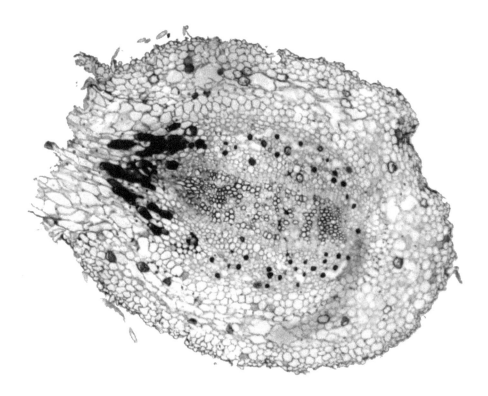

12. Examine slide 153 labeled *Ginkgo* leaf c.s. Identify **dermal layer, mesophyll, vascular tissue,** and **stomates**.

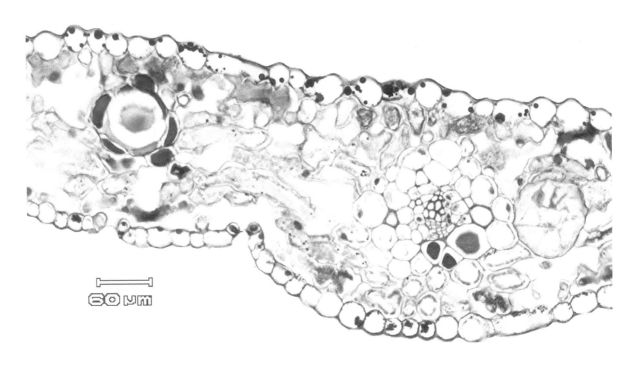

13. Examine slide 154 labeled *Ginkgo*: male strobilus l.s. Identify the **axis of the pollen cone, pollen sacs (microsporangia),** and **pollen grains.**

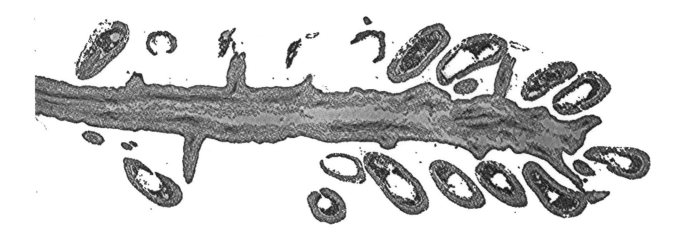

14. Examine slide 166 labeled *Ginkgo* ovule with micropyle. Identify the **collar, integument, nucellus, micropyle, pollen chamber,** and **possibly a tetrad of megaspores**.

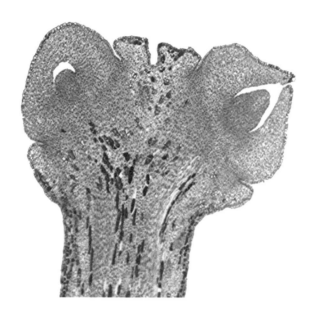

15. Examine slide 114 labeled *Ginkgo:* embryo l.s. med. Identify the **suspensor, female gametophyte, shoot apex, nucellus,** and **cotyledon primordium**.

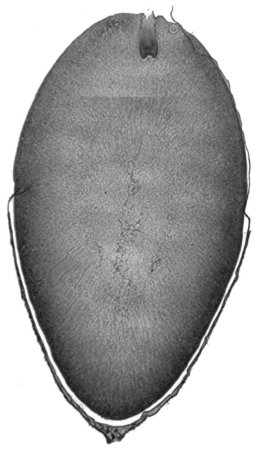

Phylum Pinophyta

The Pinophyta, also known as the Coniferophyta, are those seed plants with vascular tissue that bear cones, thus they are oftentimes referred to as the conifers (from the Latin *conus* for cone and *ferre* for bear). Their fossil record dates back to the Carboniferous Period (360-240 Ma). The group is represented by eight families with nearly 70 genera and approximately 600 species. The majority of these are trees and just a few are shrubs. They are the dominant tree species in the boreal forest ecosystem and are an important source of pulp/paper and wood products in the paper and lumber industries globally. The lay persons' view of this group is that they are evergreen; however, the conifers also include some deciduous species. The conifers further represent the oldest (9550 year old *Picea abies*, named Old Tjikko), tallest (115.5 meters tall *Sequoia sempervirens*, named Hyperion), and largest trees (11.1 meters at the base and weighing 1256 metric tons, *Sequoiadendron giganteum*, named General Sherman).

16. Examine slide 218 labeled *Pinus* first year stem. Identify the **periderm, resin canals, cortex, phloem, cambial zone, secondary xylem, pith,** and **annual ring**.

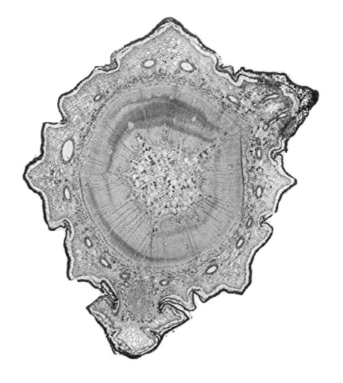

17. Examine slide 164 labeled *Pinus* older stem. Identify the **periderm, resin canals, cortex, phloem, cambial zone, secondary xylem, pith,** and **annual ring**.

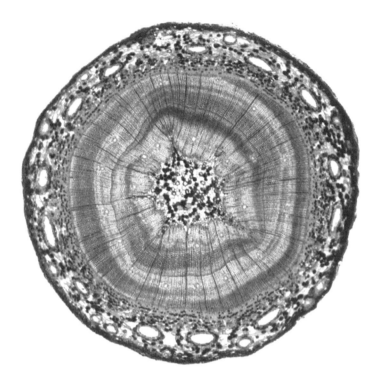

18. Examine slide 167 labeled *Pinus*: meristematic stem. Is this a cross section or a longitudinal section? Identify **pith, cortex, vascular tissue, needles,** and **lateral branches**.

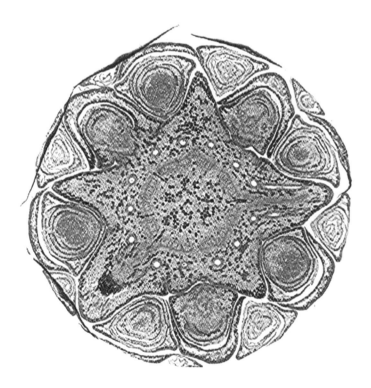

19. Examine slide 165 labeled *Pinus*: stem, cross, radial, tangential sections. Identify the **xylem, cambium, phloem,** and **cortex** in each. How does the anatomy compare in the sections? Identify which of the sections is **cross,** which is **radial,** and which is **tangential.**

20. Examine slide 168 labeled *Pinus*: stem, tip, median l.s. Identify the **shoot apical meristem, leaf primordia,** the **ground tissue,** and the **vascular tissue** of the stem.

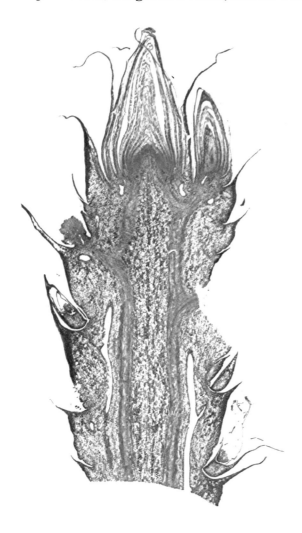

21. Examine slide 173 labeled *Pinus* root c.s. Identify the **pith, xylem, phloem, cortex,** and **epidermis.**

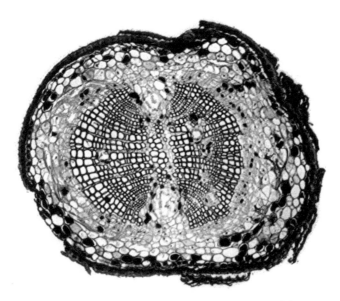

22. Examine slide 174 labeled *Pinus* old root c.s. How does the anatomy compare to slide 173?

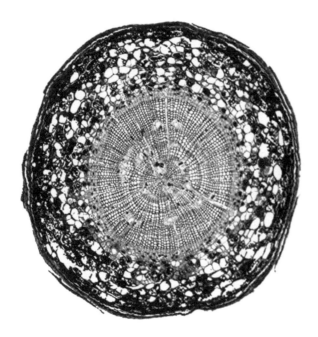

23. Examine slide 176 labeled *Pinus* root c.s. mycorrhizae. Identify the **hyphae**.

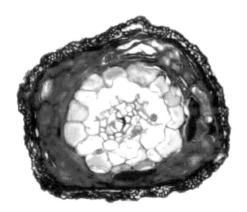

24. Examine slide 175 labeled *Pinus* root l.s., tip median. Identify the **root cap, epidermis, cortex, vascular cylinder,** and **apical meristem**.

25. Examine slide 178 labeled *Pinus* leaf c.s. Identify the **resin canals, phloem, xylem, sclerenchyma, mesophyll, endodermis,** and **stoma**.

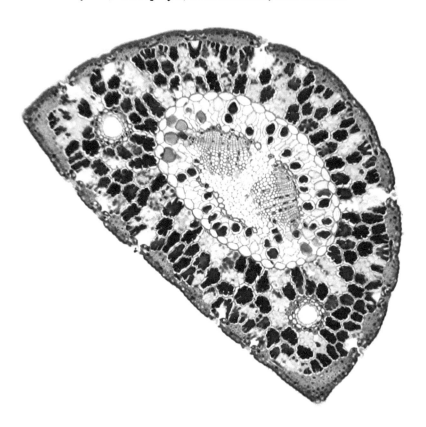

26. Examine slide 217 labeled *Pinus*: leaf c.s. (5 - needle type). Identify the **resin canals, phloem, xylem, sclerenchyma, mesophyll, endodermis,** and **stoma**. How does the anatomy compare to that of slide 178?

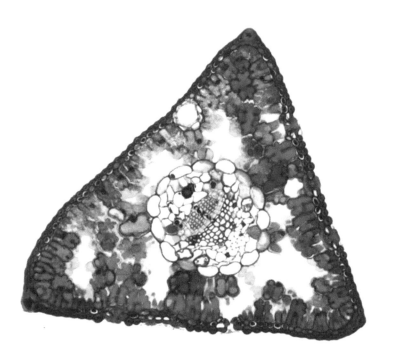

27. Examine slide 219 labeled *Pinus*: mature male strobilus median l.s. Identify **the pollen cone axis, pollen cone scale, pollen sac (microsporangium),** and **pollen grains**.

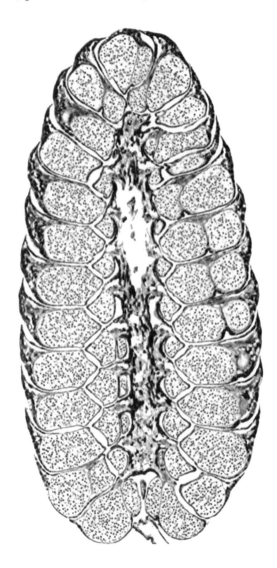

28. Examine slide 181 labeled *Pinus* mature pollen, w.m. Identify the **air bladders**.

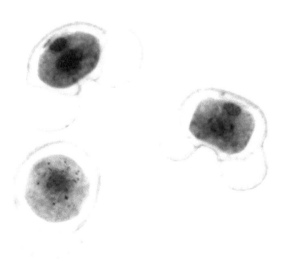

29. Examine slide 182 labeled *Pinus*: megaspore mother cell. Identify the **ovulate cone axis, bract (scale leaf), megasporangium, ovuliferous scale, micropyle,** and **pollen chamber**.

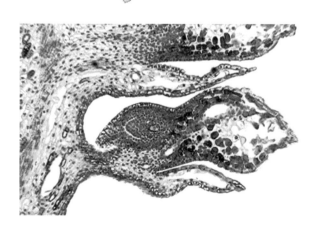

30. Examine slide 183 labeled *Pinus* proembryo (mature megagametophyte). Identify the **nucellus, pollen chamber, gametophyte, egg, archegonia,** and **archegonial chamber**.

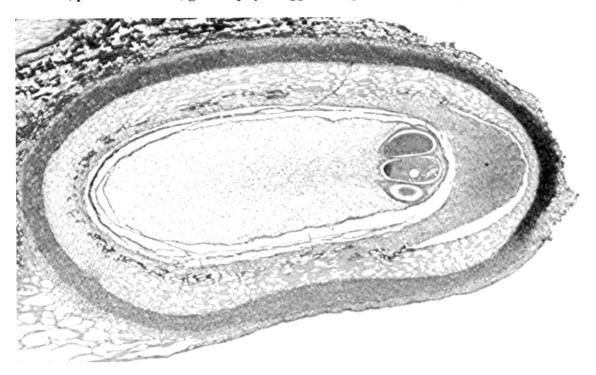

31. Examine slide 185 labeled Pinus embryo cotyledons, young. Identify the **seed coat (if present), female gametophyte (endosperm), cotyledons, shoot apex,** and **root apex**. How many cotyledons are visible in this section? How many are visible in your section?

32. Draw and diagram the life cycle of *Pinus*. Identify which phases of the life cycle are **haploid** and which are **diploid**. Where do **syngamy, meiosis,** and **mitosis** respectively occur?

Phylum Gnetophyta

The Gnetophyta is represented by three Orders, the Ephedrales, Gnetales, and Welwitschiales, each of which includes a single genus, *Ephedra, Gnetum,* and *Welwitschia* that collectively encompass less than 100 species. The Gnetophytes differ from other gymnosperms in that vessel elements are present, archegonia are absent in *Ephedra* and *Gnetum*, and sperm is non-flagellated. These characteristics represent similar character states to those within the Magnoliophyta (angiosperms or flowering plants). In the past, these 'commonalities' were believed to suggest a direct common ancestry between the Gnetophyta and the Magnoliophyta. Today, molecular data does not fully corroborate this relationship and hence the evolutionary relationship between the Gnetophyta and the Magnoliophyta remain unclear, and some of their striking similarity is due to convergent/parallel evolution.

33. Examine the image of *Ephedra viridis* below. Note the segmented and ridged stem. Which group examined previously had similar ridging and segmentation?

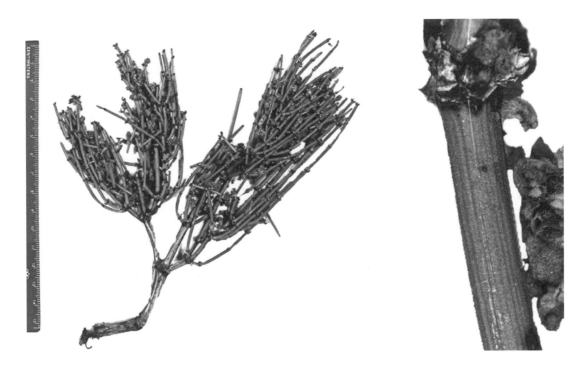

238

34. Examine slide 255 labeled *Ephedra californica* stem c.s. Identify the **pith, cortex, xylem, phloem,** and **sclerified fibers**.

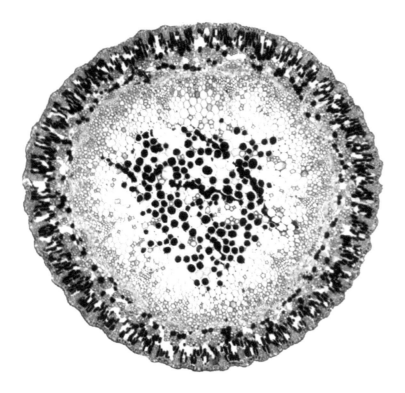

35. Examine the images of *Ephedra* ovulate strobili and microstrobili. Note the arrangement of the decussate bracts in each. Identify the **protuberant micropylar tubes** in the ovulate strobili, **dissected microsporangiophore,** and **microsporangia** in the microstrobili.

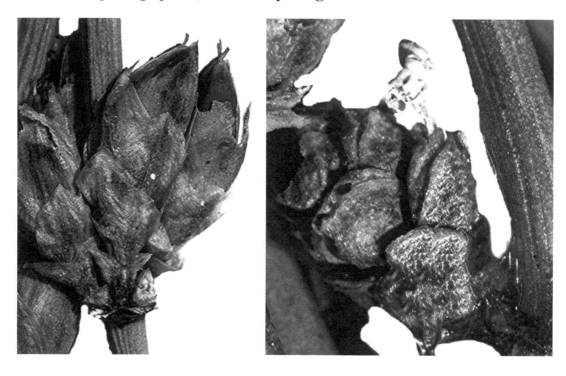

36. Examine slide 256 labeled *Ephedra* male strobilus c.s. Unlike the simple male cones of other gymnosperms these are compound. Identify the **cone axis**, **decussate bracts**, **transparent bracteoles (each encapsulate a young microstrobilus), dissected microsporangiophore, microsporangia,** and **pollen**.

37. Examine slide 258 labeled *Gnetum* leaf c.s. Identify the **epidermis, mesophyll, stomates, xylem,** and **phloem**.

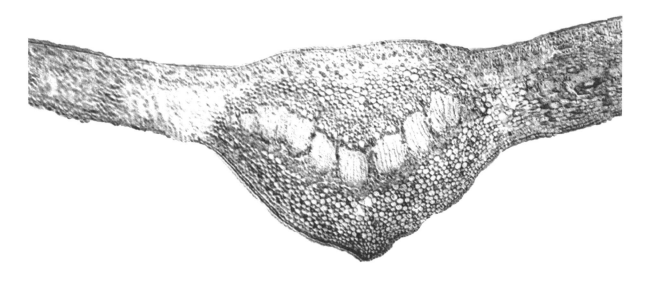

38. Examine the living demonstration material of *Welwitschia mirabilis*.

241

Readings

Andrews, H.N., Jr. 1948. *Metasequoia* and the living fossils. Missouri Botanical Bulletin 36:79-85.

Carlquist, S. 1996. Wood, bark and stem anatomy of New World species of *Gnetum*. Botanical Journal of the Linnean Society 120: 1-19.

Carlson, J.J., Farquhar, J.W., DiNucci, E., Ausserer, L., Zehnder, J., Miller, D., Berra, K., Hagerty, L., & Haskell, W.L. 2007. Safety and efficacy of a *Ginkgo biloba* - containing dietary supplement on cognitive function, quality of life, and platelet function in healthy, cognitively intact older adults. Journal of the American Dietetic Association 107: 422-432.

Chaw, S.-M., Parkinson, C.L., Cheng, Y., Vincent, T.M., & Palmer, J.D. 2000. Seed plant phylogeny inferred from all three plant genomes: Monophyly of extant gymnosperms and origin of Gnetales from conifer. Proceedings of the National Academy of Sciences 97: 4086-4091.

Just. T. 1948. Gymnosperms and the origin of angiosperms. Botanical Gazette 110: 91-103.

Ndam, N., Nkefor, J.–P., & Blackmore, P. 2001. Domestication of *Gnetum africanum* and *G. buchholzianum* (Gnetaceae), over-exploited wild forest vegetables of the Central African Region. Systematics and Geography of Plants 71: 739-745.

Pielou, E.C. 2011. The World of Northern Evergreens, 2nd ed. Cornell University Press, Ithaca.

Shaw, F.J.F. 1908. A contribution to the anatomy of *Ginkgo biloba*. New Phytologist 7: 85-92.

Stevenson, D. 1992. A formal classification of the extant cycads. Brittonia 44: 220-223.

Stockey, R.A. 1981. Some comments on the origin and evolution of conifers. Canadian Journal of Botany 59: 1932-1940.

Takaso, T. & Tomlinson, P.B. 1992. Seed cone and ovule ontogeny in *Metasequoia, Sequoia* and *Sequoiadendron* (Taxodiaceae-Coniferales). Botanical Journal of the Linnean Society 109: 15-37.

Wang, Z.-Q. 2004. A new Permian gnetalean cone as fossil evidence for supporting current molecular phylogeny. Annals of Botany 94: 281-288.

Domain Eukarya
  Supergroup Archaeplastida
    Phylum Magnoliophyta (Flowering Plants)
      Class Magnoliopsida (Dicotyledonae)
      Class Liliopsida (Monocotyledonae)

Related Terms: anthophytes, strobilus-like reproductive structures, vessels, placentation, ovary position, axile, parietal, free central, hypogynous, perigynous, epigynous, egg, synergids, polar, antipodal, globular, heart, torpedo, solitary, spike, corymb, raceme, panicle, fruits, double fertilization, flower, co-evolution, pollination biology, sepals, petals, tepals, calyx, corolla, corona, androecium, gynoecium, stamen, anther sac, filament, pistil, carpel, stigma, style, ovary, ovules, double fertilization

Angiosperms

Phylum Magnoliophyta

The angiosperms (from the Greek *angeion* for vessel and *sperma* for seed) are also known as the flowering plants. Unlike the seeds of the gymnosperms, those of the angiosperms are encapsulated, that is they are covered by a fleshy or papery/leathery layer of tissues. This monophyletic group is the most common component of the world's terrestrial flora. The group includes approximately 250,000 extant species in the Phylum Magnoliophyta that have been traditionally placed within two Classes, the Magnoliopsida (dicots) and Liliopsida (monocots). For the sake of simplicity we will adopt this classification scheme, though today pollen morphology, carpel closure through secretion or fusion, and secondary plant compounds in major part are used to divide the dicots into three groups, the paleodicots, magnoliids and eudicots. The most significant feature possessed by the group is the flower, a structure that typically possesses male and female reproductive parts. Although it is plausible that seed ferns are ancestral to flowering plants, the incomplete fossil record in conjunction with inconclusive molecular data, do not provide sufficient evidence to support this hypothesis. Darwin labeled the apparent sudden appearance of flowers in the fossil record as an abominable mystery and it remains a topic of great debate today. *Archaefructus*, which dates back to 120 Ma is considered to be the oldest known flowering plant fossil. This said, pollen attributed to flowering plants dates back to 130 Ma.

I.   Flower

The typical/classical flower is composed of four floral series that occur as whorls. These floral series are (from the outside to inside) the sepals (calyx), petals (corolla), stamens (androecium, from the Greek *andros* for male and *oikia* for house; meaning male house) and carpels (gynoecium, from the Greek *gyne* for female and *oikos* for house; meaning female house). If all four whorls are present the flower is perfect. If one or more of the series is absent the flower is imperfect. If both androecium and gynoecium are present the flower is complete. If either the androecium or gynoecium is missing the flower is incomplete. Although wind pollination is the most primitive form of pollination per se, wind pollination represents an advanced mode of pollination in the flowering plants. In this group beetle pollination is primitive. Both the diversification and explosion in numbers of the angiosperms was a consequence of co-evolution that occurred between the pollinating agent and respective flowering plant.

II.   Seed

The seed is a little plant (embryo) in a box (pericarp=seed coat) with its lunch (endosperm). The embryos have either one or two seed leaves (cotyledons) and in this respect are monocotyledonous or dicotyledonous. The cotyledon(s) wither shortly after their emergence and are replaced by true leaves that provide the necessary photosynthates for survival. These two groups are generally referred to as monocots, paleodicots, and eudicots, respectively. In more modern classification schemes the angiosperms are subdivided into the monocots, dicots, and eudicots. For the sake of simplicity, we will collectively refer to the angiosperms as having only two subgroupings (i.e. the monocots and dicots).

III.   Fruit

A fruit is the ripened ovary or ovaries and in some cases also includes non-ovarian tissues that generally results from sexual reproduction. Unfortunately, sometimes this process is more complicated in that fruits may be produced without the transfer of pollen to the stigmatic surface of the carpel, but we will not discuss these special cases at this time. Ultimately fruit type is based on the anatomy of the pericarp (the exocarp, mesocarp and endocarp collectively), fertilized ovule and sometimes associated non-ovarian tissue.

Fleshy fruits have a fleshy/soft pericarp and may be simple or compound. Simple fruits are derived from a single ovary and are a combination of pericarp and the seed within (e.g. peach and grape). Compound fruits are formed from the fusion of many ovaries and possibly accessory tissue. These compound fruit types include: aggregate, accessory, and multiple fruits. An aggregate fruit results from the fusion of multiple ovaries from a single flower (e.g. blackberry). A multiple fruit results when ovaries of multiple flowers fuse to form a fruit (e.g. pineapple). Accessory fruits are derived from a single ovary or many ovaries in which non-ovarian tissues is also incorporated into the accessory fruit (e.g. strawberry).

Specialized types of simple and compound fruits are the dry fruits. Unlike the fleshy fruits, dry fruits have a woody/papery or leathery pericarp. Dry fruits include: indehiscent dry fruits and dehiscent dry fruits. Indehiscent dry fruits do not open along a predetermined line of dehiscence (e.g. samara of a maple). Dehiscent dry fruits open along a predetermined line of dehiscence (e.g. pod of a legume). Dry fruits in addition to being indehiscent or dehiscent are either simple (e.g. silique of a mustard) or compound (e.g. glans of an oak).

IV.   Anatomy

The tissue system that encapsulates the plant is the dermal tissue system and consists of parenchyma cells. Internal to the dermal tissue system is the ground tissue system. The ground tissue system can be composed of three main tissue types namely, parenchyma tissue (parenchyma cells), collenchyma tissue (collenchymas cells), and sclerenchyma tissue (fibers and sclereids). The vasculature tissue system contains both phloem and xylem tissue. The former consists of sieve-tube member cells, companion cells, and parenchyma cells, and the latter may consist of tracheids, vessel elements, parenchyma cells, and fibers. The arrangement of the xylem and phloem in the stems and roots (stelar anatomy) can be used to further support the subdivision of the angiosperms into two groups, that is, the monocots and dicots. In stems the monocots generally exhibit an atactostele (a scattered arrangement of vascular bundles of xylem and phloem intermixed within in the ground tissue), whereas the dicots generally exhibit a eustele (a single distinct ring of vascular bundles of xylem and phloem within the ground tissue). In roots the monocots exhibit a polyarch arrangement of xylem (a star-like arrangement that surrounds a pith) whereas dicots generally exhibit an actinostele (a star-like arrangement of xylem that does not surround a pith). Tissues with only one cell type are simple. Those with two or more cell types are compound/complex.

Class Magnoliopsida – herbaceous

1. Examine slide 226 labeled *Helianthus*: root c.s. Identify the **xylem, phloem, endodermis, casparian strip, epidermis, cortex, pericycle,** and **root hairs**. Where do root hairs arise? Where do lateral roots arise? What type of stelar anatomy is exhibited in a *Helianthus* root?

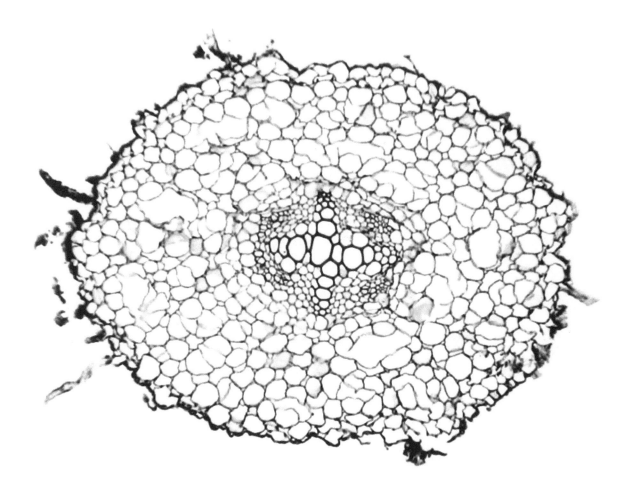

2. Draw and label a tissue diagram for the *Helianthus* root illustrated above.

3. Examine slide 228 labeled *Helianthus*: old stem x.s. and slide 229 labeled development of interfascicular cambium - *Helianthus*. Note that both the fascicular and interfascicular cambial regions have become active. Identify **fascicular** and **interfascicular cambium, pith, cortex, epidermis, fibers, xylem,** and **phloem**. What type of stelar anatomy is exhibited in a *Helianthus* stem?

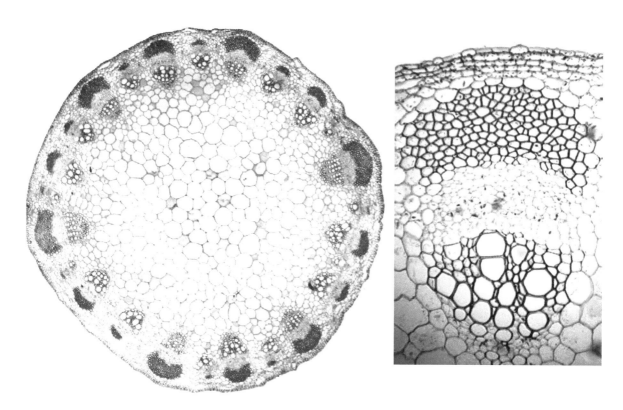

4. Examine slide 247 labeled *Helianthus* stem tip l.s. (some slides are labeled *Helianthus* stem tip median l.s.). Identify **leaf primordia, leaf hairs, epidermis, pith, vascular tissue, cortex, and meristematic cells**. Where is cellular differentiation occurring?

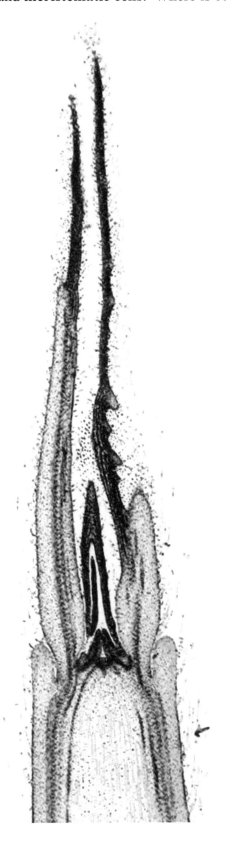

5. Examine slide 230 labeled *Helianthus* stem l.s. Identify the **vessels elements** with open ends, **tracheids** with tapered ends, **sieve tube cells, sieve plates, companion cells, fibres,** and **parenchyma cells**. How can you differentiate between a companion cell and a sieve tube cell?

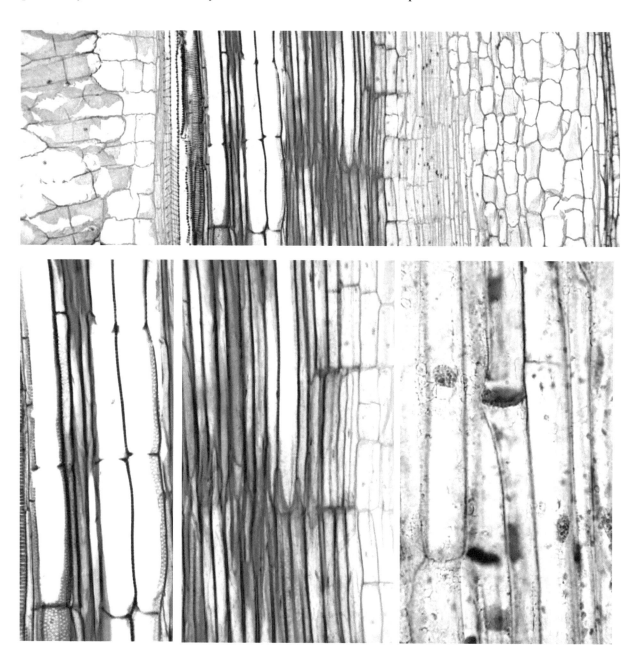

6. Examine slide 231 labeled *Helianthus* macerated stem w.m. Identify cell types within the maceration. Why are these cell types present and not others?

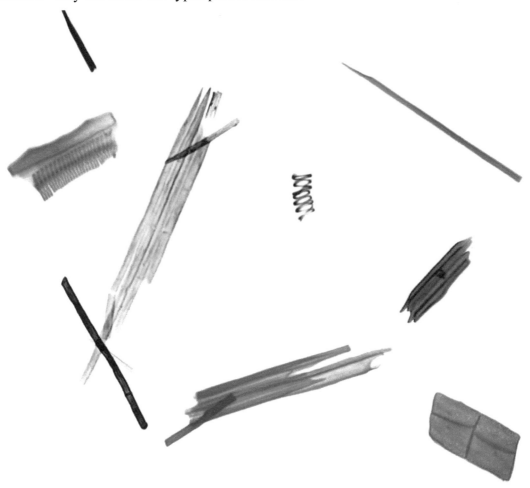

7. Examine slide 232 labeled *Helianthus*: leaf x.s. Identify **stomata, vascular bundles, epidermis, mid-vein, lateral vein, palisade** and **spongy mesophylls**. What aspects of internal anatomy define this section to be from a dicot?

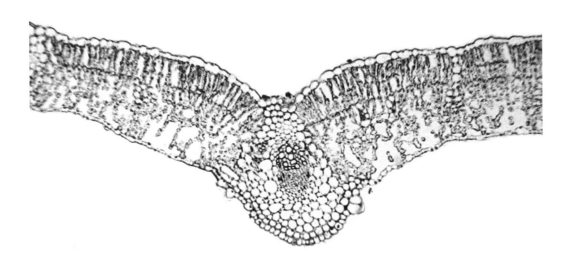

8. Examine slide 235 labeled *Ranunculus acris*: young and mature roots c.s. The innermost layer of the cortex is compactly arranged and lacks air spaces. This layer, the endodermis, is characterized by the presence of Casparian strips in its anticlinal walls (the walls perpendicular to the surface of the root). The Casparian strip is a band-like portion of the primary wall that is impregnated with a fatty substance called suberin and is sometimes lignified. The plasma membranes of the endodermal cells are quite firmly attached to the Casparian strips. The endodermis is compact and the Casparian strips are impermeable to water and ions. Thus, all substances entering and leaving the vascular system must pass through the protoplasts of the endodermal cells. This is accomplished either by crossing the plasma membranes of these cells or by passing through the numerous plasmodesmata connecting the endodermal cells with the protoplasts of neighboring cells of the cortex and vascular cylinder. Note that the pericycle is internal to the endodermis. In the young root the pericycle is composed of parenchyma cells. In most seed plants lateral roots arise in the pericycle. In plants undergoing secondary growth, the pericycle contributes to the vascular cambium opposite the protoxylem and generally gives rise to the first cork cambium. Pericycle often proliferates, that is, it gives rise to more pericycle.

Indentify **pericycle, endodermis, the casparian strip, xylem,** and **phloem.** What type of stelar anatomy is this?

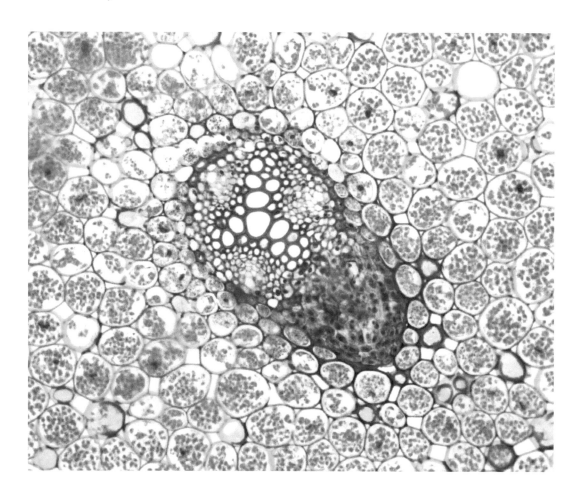

9. Examine slide 236 labeled *Phaseolus* (bean): root tip, med. l.s. Identify the **root cap, apical meristem, vascular tissue, ground tissue,** and **dermal tissue**.

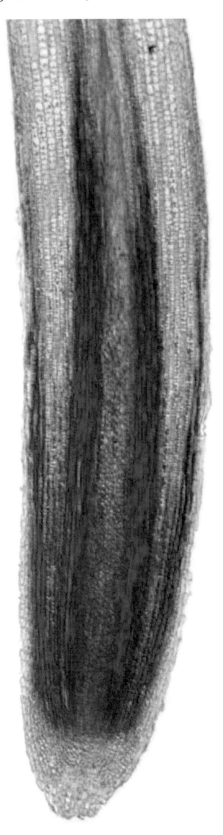

10. Examine slide 237 labeled *Phaseolus* hypocotyl, x.s. The hypocotyl is a portion of the seedling situated between the cotyledons and the radicle (embryonic root). This is in contrast to the epicotyl, that is, the upper portion of a seedling above the cotyledons and below the next leaf or leaves. Identify the tissue and cell types.

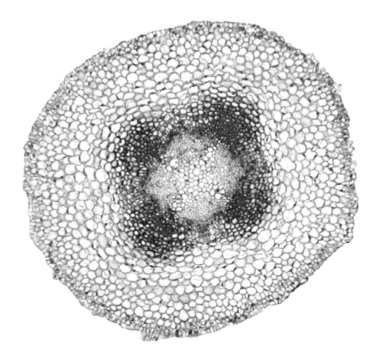

11. Examine slide 238 labeled *Phaseolus vulgaris* root tip, serial x.s. Identify the tissue and cell types.

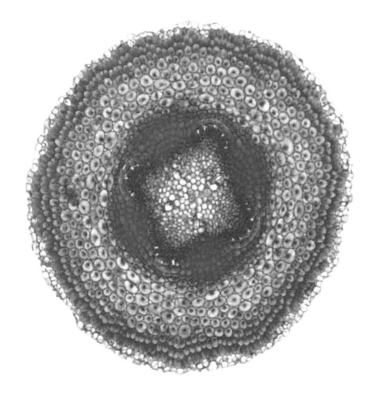

Class Magnoliopsida – woody

12. Examine slide 234 labeled *Quercus*: older stem crt. Note **secondary growth, the pith, pith rays, phloem,** and **phloem fibres**. The hardwoods (*Quercus* is an oak tree) are usually subdivided into two major groups on the basis of the presence or absence of rings of pores (early wood vessels) in cross sections. In **ring porous wood** the early wood vessels are appreciably larger than those of the late wood. In **diffuse porous wood** the early wood vessels are not larger or are only slightly larger than those of late wood. *Quercus* is ring porous.

Cross section

Radial section   **Tangential section**

13. Examine slide 233 labeled *Platanus* stem x.s. Note **secondary growth, the pith, pith rays, phloem,** and **phloem fibres**. Is *Platanus* **ring porous** or **diffuse porous**. Explain?

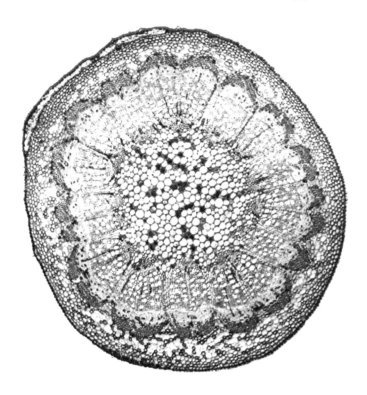

Class Liliopsida

14. Examine slide 239 labeled *Zea mays*: root tip x.s. with hairs. Identify respective cell types. What type of stelar anatomy does this root exhibit?   Which cells give rise to the root hairs?

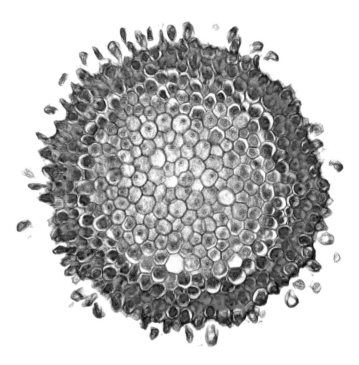

15. Examine slide 240 labeled *Zea mays*: root tip serial x.s. The large circular objects are vessels at different stages of development. Identify the **endodermis, pericycle, cortex, pith,** and **point of origin of the lateral roots**. What type of stelar anatomy does this root exhibit?

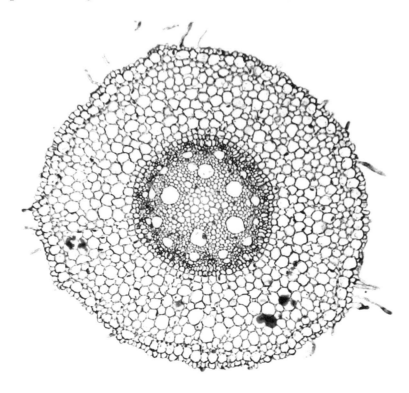

16. Examine slide 246 labeled *Zea mays*: branch root origin. Which cells give rise to lateral roots?

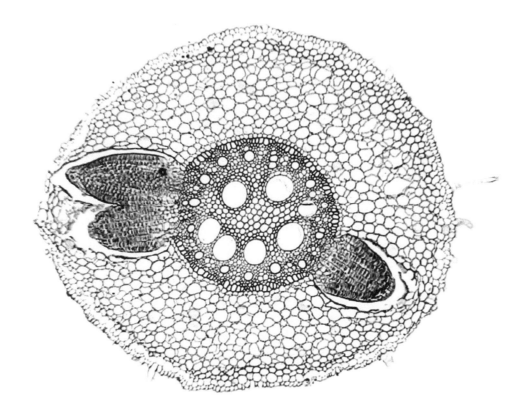

17. Examine slide 249 labeled *Zea mays*: mature stem c.s. Identify the **vascular bundles** and describe their arrangement.

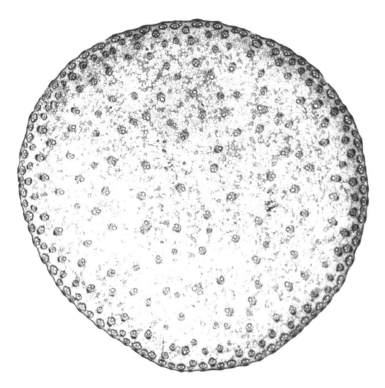

18. Examine slide 250 labeled *Zea mays*: l.s. young. Identify respective cell types. Why is the vasculature far less conspicuous than that of a dicot?

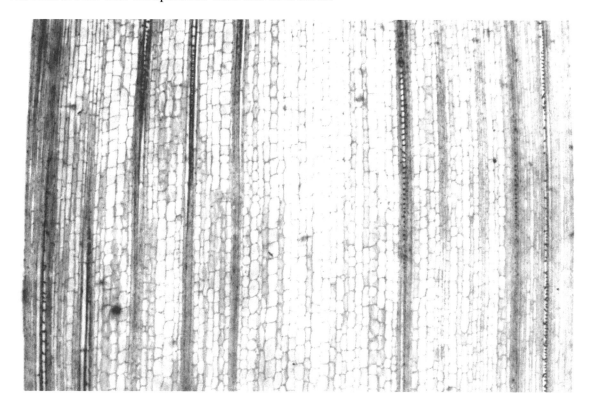

19. Examine slide 243 labeled *Zea mays*: macerated stem w.m. Identify cell types within the maceration. Why are these cell types present and not others?

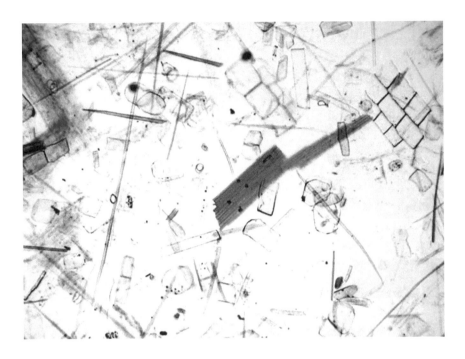

20. Examine slide 241 labeled *Zea mays*: node: c.s. Identify the **vascular bundles** and **leaf traces**. What type of stelar anatomy does this stem exhibit?

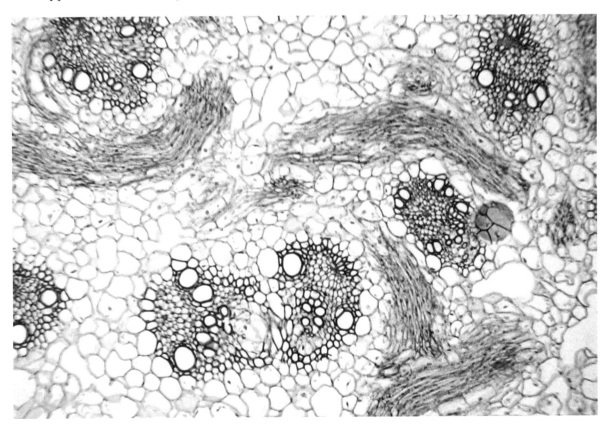

21. Examine slide 242 labeled *Zea mays*: node l.s. Note the arrangement of the various tissues types. Identify **leaf traces** and **leaf gaps**.

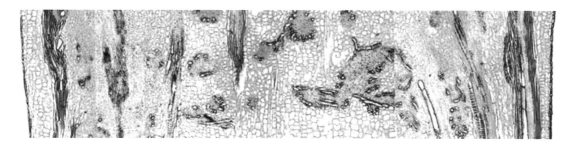

22. Examine slide 244 labeled *Poa pratense*: leaves, folded and open, c.s. Identify the **xylem, phloem, fibres, cells of the mesophyll, stomates,** and **bulliform cells**. Bulliform cells are large epidermal cells which occur in longitudinal rows in grass leaves and are believed to be involved in the mechanisms of rolling and unrolling.

Microgametogenesis

23. Examine the slide labeled *Lilium*: young bud l.s. (Triarch slide no. 17-7). Note the **peduncle**, **stigma, style, ovary, anther filaments, anther sacs, petals,** and **sepals**.

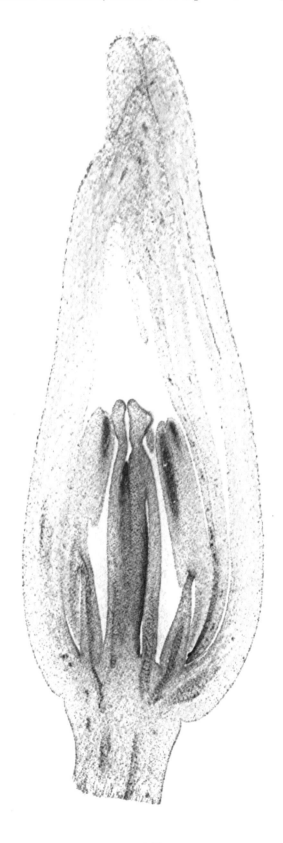

24. Examine the slide labeled *Lilium*: bud c.s., early anthers (Triarch slide no. 18-1). Note the **sepals, petals, anther sacs, ovary**, and **carpels**.

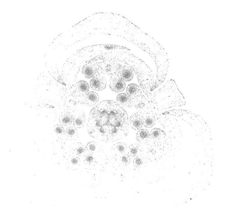

25. Examine the slide labeled *Lilium*: anthers c.s., synisezis (Triarch slide no. 18-2). Synisezis is the massing of chromatin of the nucleus preceding division.

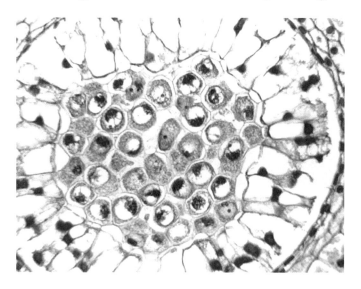

26. Examine the slide labeled *Lilium*: anthers c.s., early prophase (Triarch slide no. 18-3).

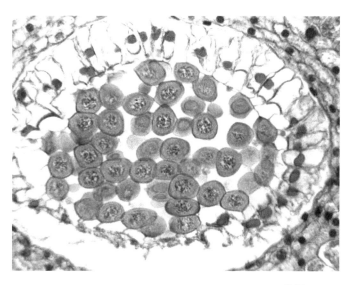

27. Examine the slide labeled *Lilium*: late prophase (Triarch slide no. 18-4).

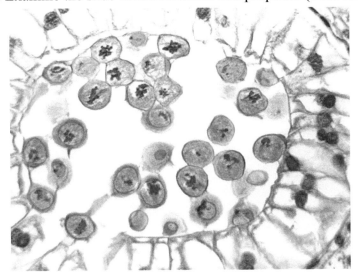

28. Examine the slide labeled *Lilium* anthers c.s., first division (Triarch slide no. 18-5).

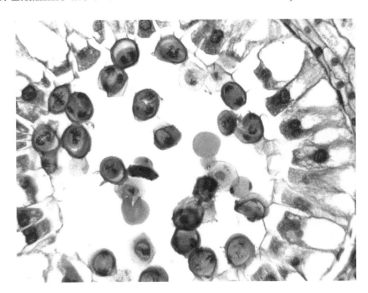

29. Examine the slide labeled *Lilium*: anthers c.s., second division (Triarch slide no. 18-6).

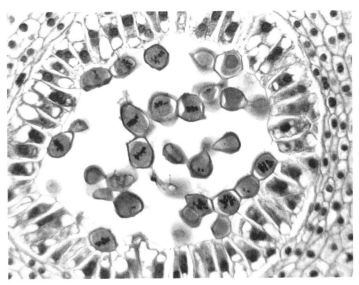

30. Examine the slide labeled *Lilium*: anthers c.s., pollen tetrads (Triarch slide no. 18-7).

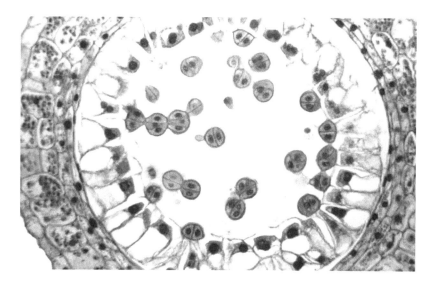

31. Examine the slide labeled *Lilium*: 1-celled microspores (Triarch slide no. 18-8).

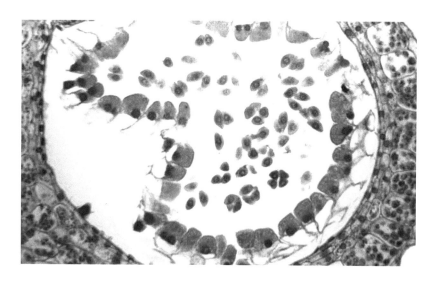

32. Examine the slide labeled *Lilium*: mature anthers (Triarch slide no. 18-9). The haploid microspore nucleus has divided mitotically to produce the tube nucleus and the generative nucleus. The pollen grain is dispersed in this condition.

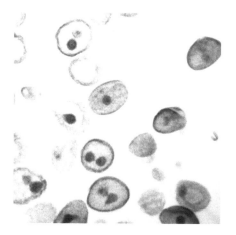

33. Examine the slide labeled *Lilium*: stigma and pollen tubes (Triarch slide no. 18-10). The generative nucleus has now divided to produce two sperm.

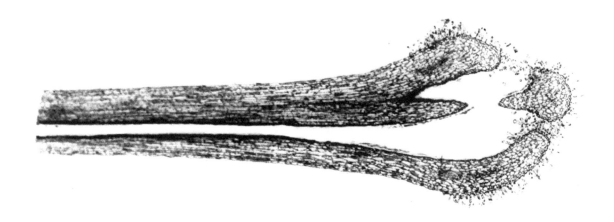

Megagametogenesis

Megagametophyte developed in the angiosperms varies. Variation exists in (1) the number of meiotic products involved in the formation of the megagametophyte (one, two, or four); (2) the number of cell divisions noted in development up to the stage when the megagametophyte is ready for fertilization; (3) the number of antipodals; (4) the ploidy level of the polar nuclei; and (5) the position of the functional megaspore in the megaspore tetrad. In spite of this developmental variation, two cells, the egg and the central cell are always present.

34. Examine the slide labeled *Lilium*: ovule megasporocyte, integuments complete (Triarch slide no. 19-2). Note the megasporocyte. Will the megasporocyte undergo **meiosis** or **mitosis**? Is the megasporocyte **haploid** or **diploid**?

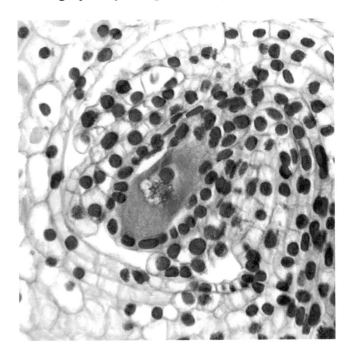

35. Examine the slide labeled *Lilium*: ovule first division (Triarch slide no. 19-3).

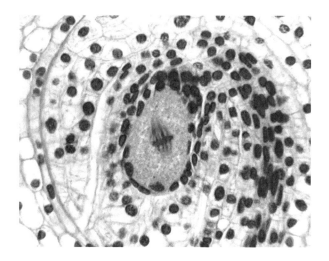

36. Examine the slide labeled *Lilium*: ovule binucleate (Triarch slide no. 19-4b) - division I.

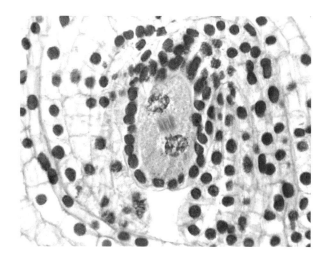

37. Examine the slide labeled *Lilium*: ovule second division (Triarch slide no. 19-5b) - division II.

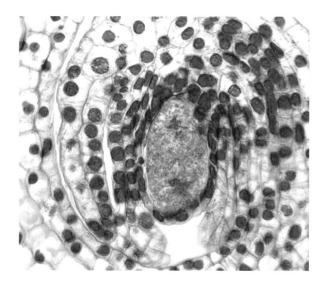

38. Examine the slide labeled *Lilium*: ovule megaspores (Triarch slide no. 19-6b) - completion of meiotic divisions - division II, 4-nucleate stage.

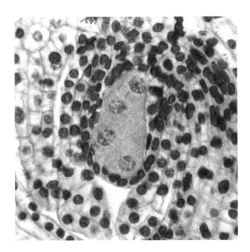

39. Examine the slide labeled *Lilium*: ovule migrating nuclei (Triarch slide no. 19-7b) – migration of 3 nuclei to chalazal end.

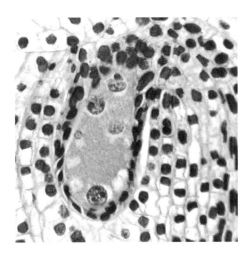

40. Examine the slide labeled *Lilium*: ovule third division (Triarch slide no. 19-8b) - 2 mitotic nuclei - division III, fusion of 3 nuclei.

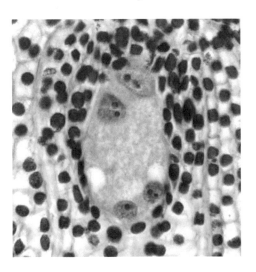

41. Examine the slide labeled *Lilium*: ovule, second 4-nucleate stage (Triarch slide no. 19-9a) - division IV.

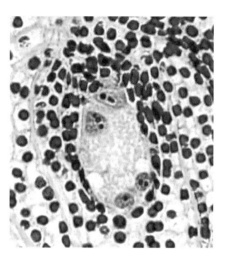

42. Examine the slide labeled *Lilium*: ovule, fourth division (Triarch slide no. 19-10b) - division IV.

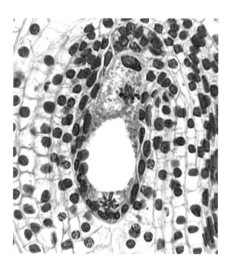

43. Examine the slide labeled *Lilium*: ovule, mature embryo sac (Triarch slide no. 19-11b) - division V.

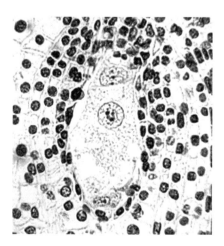

44. Examine the slide labeled *Lilium*: ovule, fertilization, egg-fusion (Triarch slide no. 19-12c-1).

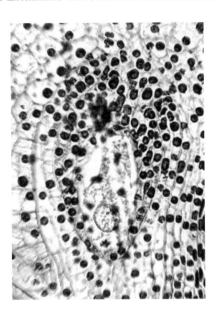

Embryo development

45. Examine slide 245 labeled *Capsella*: campylotropous ovule: A campylotropous ovule is one in which the funiculus is attached near the equator of the ovule. Note the young pollen grains and ovules.

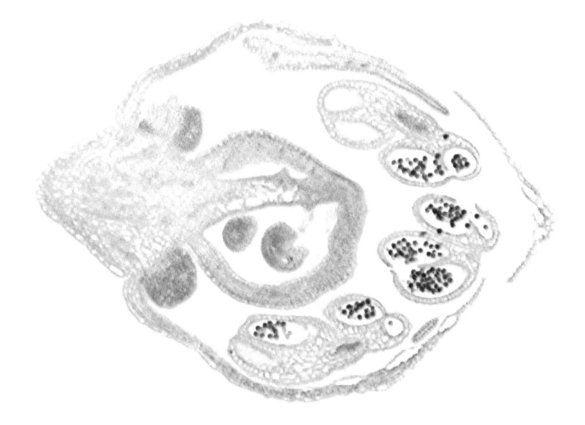

46. Examine the slide labeled *Lilium:* early embryo (Triarch slide no. 19-14).

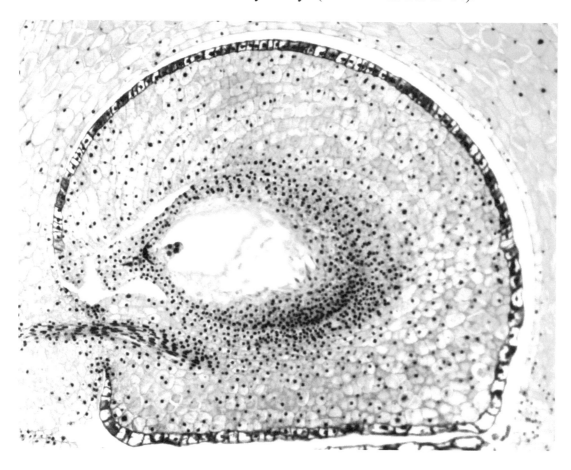

47. Examine slide 221 labeled *Capsella*: early cotyledons l.s. Identify the **suspensor with large basal cell, early embryo, endosperm, nucellus,** and **antipodal tissue**. This stage of embryo development is typically referred to as the **globular stage**.

48. Examine slide 223 labeled *Capsella*: early cotyledons l.s. non-median. Identify the **suspensor, large basal cell, endosperm, antipodal tissue, protoderm, embryonic root tip, nucellus,** and **emerging cotyledons**. This stage of embryo development is commonly referred to as the **heart stage**. Is *Capsella* a **monocot** or **dicot**? Explain.

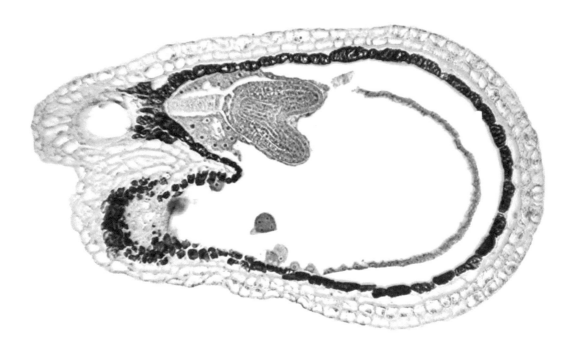

49. Examine slide 223 labeled *Capsella*: early cotyledons l.s. non-median. Identify the **suspensor, large basal cell, endosperm, antipodal tissue, protoderm, embryonic root tip, nucellus,** and **expanding cotyledons**. This stage of embryo development is known as the torpedo.

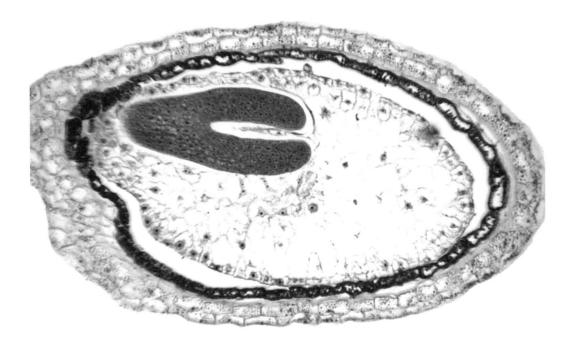

50. Examine slide 224 labeled *Capsella*: bending cotyledons l.s. Identify the **basal cell**, **antipodal tissue, protoderm, embryonic root tip, endosperm, ground meristem, procambium, protoderm,** and **bending cotyledons.** This stage of development is just past the torpedo stage that is that stage where the cotyledons have expanded but have not yet begun to bend.

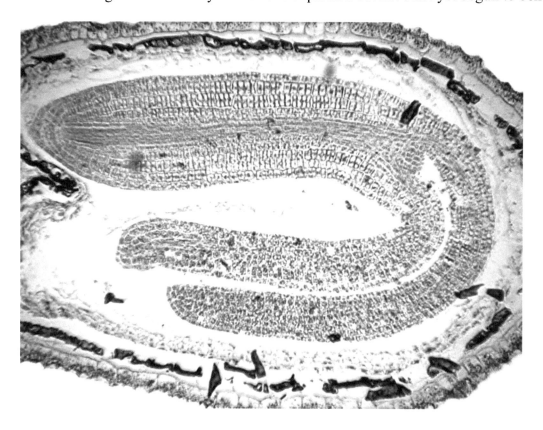

51. Examine slide 225 labeled *Capsella*: mature embryo l.s. Identify the **endosperm, nucellus, root cap, shoot apical meristem, cotyledons, seed coat,** and **funiculus.**

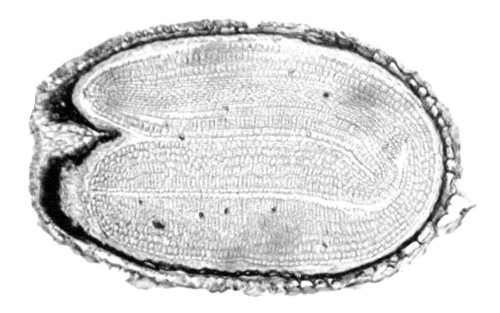

52. Examine slide 19-15 labeled *Lilium*: mature seed c.s.

53. Examine slide 252 labeled *Zea mays*: embryo l.s. (root). Identify the **pericarp** (the fruit wall which develops from the mature ovary wall), **endosperm**, **scutellum** (a single cotyledon of a grass embryo, specialized for the absorption of endosperm), **coleoptile** (the sheath enclosing the apical meristem and leaf primordia of the monocot embryo which is often interpreted as the first leaf), **shoot apex**, **coleorhiza** (the sheath enclosing the radicle in the monocot embryo), and **radicle** (the embryonic root).

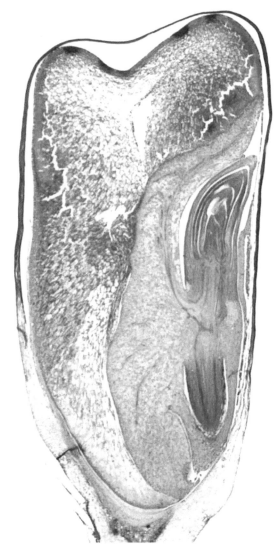

Flower

54. Use the sectioned image of a representative member of the Order Plasticales to identify **sepals**, **petals**, **perianth**, **calyx**, **corolla**, **corona**, **androecium**, **gynoecium**, **stamens**, **anthers**, **filaments, carpels, stigma**, **style**, **ovary,** and **ovules**. Is the ovary **inferior** or **superior**? Explain. Is this flower a **monocot** or a **dicot**? Explain.

55. Use the sectioned image of *Aquilegia canadensis* to identify **sepals**, **petals**, **calyx**, **corolla**, **corona**, **androecium**, **gynoecium**, **stamens**, **anthers**, **filaments, carpels, stigma**, **style, ovary,** and **ovules**. Is the ovary **inferior** or **superior**? Explain. Is this flower a **monocot** or a **dicot**? Explain.

56. Use the image of *Narcissus pseudonarcissus* below to identify **sepals**, **petals**, **calyx**, **corolla**, **corona**, **androecium**, **gynoecium**, **stamens, anthers, filaments, carpels, stigma, style, ovary, ovules, perianth,** and **tepals**. Is the ovary **inferior** or **superior**? Explain. Is this flower a **monocot** or a **dicot**? Explain.

57. Diagram the typical life cycle of a flowering plant that is the *Lilium* type. Fully detail sporogenesis and gametogenesis. Identify **haploid, diploid,** and **triploid** phases throughout the life cycle. Also include where **syngamy, meiosis,** and **mitosis** occur.

Fruit

Recall that a fruit is the ripened ovary or ovaries and in some cases also includes non-ovarian tissues of a flowering plant that generally results from sexual reproduction. Use the terms below to complete this section of the laboratory exercise.

Definition of Terms:

Seed - matured ovule

Fruit - a matured ovary and its contents, together with other floral parts that may sometimes adhere to it

The ovary wall (pericarp) consists of three layers in fruits

(a) Exocarp - outer layer (skin)

(b) Mesocarp - middle layer (fleshy)

(c) Endocarp - inner layer

The fruit is oftentimes an important diagnostic feature of a family or genus. There are two ways of approaching the classification and naming of fruits

(a) **Morphological -** fruits are distinguished by their fleshiness dehiscence, carpel number, etc.

(b) **Taxonomic -** a particular fruit is given a name because of the family to which it belongs, with little regard to details of its structure

SIMPLE FRUIT - fruits derived from the ovary of a solitary pistil in a single flower

   (a) **Dry Indehiscent Fruit** - dry fruits that do not split open at maturity

      ACHENE - a one-seeded, fruit with the seed attached to the fruit wall at one point only

      CAPSULE - dry fruit derived from a two or more loculed ovary

      CARYOPSIS / GRAIN - one seeded fruit with the a seed coat adnate to the fruit wall

      NUT - one-seeded fruit with a hard pericarp

      SAMARA - a winged dry fruit

   (b) **Dry Dehiscent Fruit -** dry fruits that open at maturity

      FOLLICLE - a dry fruit derived from one carpel that splits along a suture

      LEGUME - a dry dehiscent fruit derived from one carpel that splits along two sutures

   (c) **Capsule -** based on the type of dehiscence

      ACROCIDAL CAPSULE - one that dehisces through terminal slits or fissures

      ANOMALICIDAL or RUPTURING CAPSULE - one that dehisces irregularly

      BASICIDAL CAPSULE - one that dehisces through basal slits or fissures

      CIRCUMSCISSLE CAPSULE / PYXIS - capsule that dehisces circumferentially

      OPERCULATE CAPSULE - one that dehisces through pores, each of which is covered by a flap, cap or lid

      PORICIDAL CAPSULE - one that dehisces through pores

   (d) **Fleshy Fruit -** fleshy ovary

      BERRY - fleshy fruit with a succulent pericarp

      DRUPE - a fleshy fruit with a stony endocarp

      HESPERIDIUM - a thick skinned septate berry with the bulk of the fruit derived from glandular hairs

      PEPO - a berry with a leathery non-septate rind

COMPOUND FRUIT - fruits derived from the fusion of many ovaries and possibly accessory tissue

(a) **Accessory Fruit -** fruit derived from simple or compound ovaries and some non-ovarian tissue

    BUR - cypsela enclosed in a dry involucre

    HIP - an aggregation of achenes surrounded by an urn-like receptacle

    POME - a berry like fruit, adanate to a fleshy receptacle

    PSEUDOCARP - an aggregation of achenes embedded in a fleshy receptacle

(b) **Aggregate Fruit -** group of separate fruits developed from the carpels of one flower

    ACHENECETUM - an aggregation of achenes

    BACCAVETUM / ETAERIO - an aggregation of berries

    SAMARACETUM - an aggregation of samaras

(c) **Multiple Fruit -** fruits on a common axis that are usually coalesced (forming one mass) and derived from the ovaries of several flowers

    SYCONIUM - a vase-like structure with an opening at the apex and an interior wall line with flowers.

    SOROSIS - A fruit derived from the consolidation of many flowers with their receptacles

58. What is the difference between a fruit and a vegetable?

59. List five common fruits and five common vegetables.

	Fruit	Vegetable
1.		
2.		
3.		
4.		
5.		

60. Label the exocarp, mesocarp, and endocarp.

61. How many carpels make up the pistil of the flower from which this fruit was derived?

62. What type of placentation occurs in the ovary of the flower from which this fruit was derived?

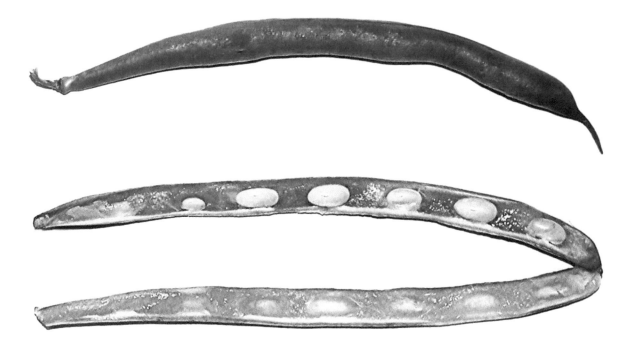

63. Use the key provided by your instructor to determine the fruit types on demonstration.

Readings

Bold, H.C., Alexopoulos, C.J., & Delevoryas, T. 1987. Morphology of plants and fungi 5th ed. HarperCollins Publishers College Division. New York, NY, USA.

Cronquist, A. 1988. The evolution and classification of flowering plants. New York Botanical Garden. New York, NY, USA.

Cronquist, A. 1981. An integrated system of classification of flowering plants. Columbia University Press, New York, NY, USA.

Esau, K. 1977. Anatomy of seed plants. 2th ed., John Wiley & Sons. New York, NY, USA.

Evert, R.F. & Eichhorn, S.E. 2013. Raven Biology of Plants 8th ed., W.H. Freeman & Company Publishers. New York, NY, USA.

Harris, J.G. & Harris, M.W. 2001. Plant identification Terminology: An illustrated glossary 2nd ed. Spring Lake Publishing, Spring Lake, UT, USA.

Hutchinson, J. 1969. Evolution and phylogeny of flowering plants. Academic Press. London, UK

Judd, W.S., Campbell, C.S., Kellogg, E.A., Stevens, P.F, & Donoghue, M.J. 2007. Plant systematics: A Phylogenetic Approach, 3rd ed., Sinauer Associates, Inc. Sunderland, MA, USA.

Masuseth, J.D. 1988. Plant anatomy, The Benjamin/Cummings Publishing Company, Inc. Menlo Park, CA, USA.

Payne, J.P., Jr. 1977. Vascular plant families. Mad River Press. Eureka, CA, USA.

Soltis, P.S. & Soltis, D.E. 2004. The origin and diversification of angiosperms. American Journal of Botany 91: 1614–1626.

Stebbins, G.L., Jr. 1950. Variation and evolution in plants. Columbia University Press, New York, NY, USA

Takhtajan, A. 1997. The diversity and classification of flowering plants. Columbia University Press, New York, NY, USA.

CPSIA information can be obtained
at www.ICGtesting.com
Printed in the USA
BVHW021116191218
535594BV00003B/5/P